The
BIG ONE

THE
BIG ONE

THE TRUE STORY OF AN EPIC SEARCH
TO FIND A MISSING SMALL PLANE
LOST FOR YEARS

Bruce Gallaher

SANTA FE

© 2013 by Bruce Gallaher
All Rights Reserved
Maps by Lori Johnson

No part of this book may be reproduced in any form or by any electronic or mechanical means including information storage and retrieval systems without permission in writing from the publisher, except by a reviewer who may quote brief passages in a review.

Sunstone books may be purchased for educational, business, or sales promotional use. For information please write: Special Markets Department, Sunstone Press, P.O. Box 2321, Santa Fe, New Mexico 87504-2321.

Book and Cover design › Vicki Ahl
Body typeface › ITC Benguiat Std
Printed on acid-free paper ∞

Library of Congress Cataloging-in-Publication Data

Gallaher, B. M., 1950–2013
   The big one : the true story of an epic search to find a missing small plane lost for years / by Bruce Gallaher.
       pages cm
   ISBN 978-0-86534-947-6 (softcover : alk. paper)
   1. Aircraft accidents--Investigation--New Mexico. 2. Search and rescue operations--New Mexico. 3. Private planes--New Mexico. I. Title.
   TL553.525.N6G35 2013
   363.12'40978956--dc23
                                                    2013016754

WWW.SUNSTONEPRESS.COM
SUNSTONE PRESS / POST OFFICE BOX 2321 / SANTA FE, NM 87504-2321 /USA
(505) 988-4418 / ORDERS ONLY (800) 243-5644 / FAX (505) 988-1025

The most beautiful thing we can experience is the mysterious.
—Albert Einstein

Prologue

SATURDAY, AUGUST 4TH, 1973

We are lost in the wilderness, yet there is no panic and we continue on. By all rights we should turn back.

For every two feet forward we drop one. The light is dim and the mist is becoming rain. Spruce and Engelmann Fir trees are spaced 10 to 20 feet apart and we can walk in the spaces, save for stepping over the occasional downed trees. In this wilderness forest one can see ahead maybe 50 feet. There is little talking between us. We have dropped for about 20 minutes now and there still is no sign of a trail. Weary, we move ahead almost mechanically until we stop.

At the edge of our view we see something that looks totally out of place—something metallic. In a few more steps I realize that the object is an airplane! On the uphill side closest to us, one end of the fuselage is hung in the trees about six feet off the ground, and we can't see underneath the plane from our view. Janie is the first to approach the plane. I'm not sure if it was the tone of her voice or the words themselves. But I shuttered when Janie turned and softly said to me "Bruce, there are bodies here."

We hike out the next day carrying identifying items from the aircraft. I call a pilot friend who participates in searches. He listens to what we found and the line goes silent.

"I think this is the Big One," he finally says.

"The Big One?"

"We looked all over the place for this and never found it. Mystery plane. It's been missing for about five years now."

Introduction

2008

My destination was one of the many small towns in south central Texas that appear every twenty miles or so between stretches of pasture. I had traveled to reconnect with my past, though I'd never been here before.

I came to see Sara. Sara was one of three women who lost their husbands in the plane crash. We had met 35 years ago for all of one hour, yet, oddly, I felt I knew her well. I had been linked forever with Sara since our discovery of the plane crash. Thirty five years passed without any contact between us, as we both redirected our lives.

Ordinarily, it would have been natural for this event to take its place in time alongside numerous other interesting things buried in the depths of long term memory. However, for some reason I just couldn't let go of this story. It continually drew me back as if it was tethered. At one point, several years passed without my consciously thinking about it. But then a flash of something would reconnect the link.

It took a long time for me to get around to writing this story. Soon after the news hit of the plane's discovery, I left town to start graduate school and my life became busy just keeping afloat with my studies. Then came a career, fun, and family. At one point, several years passed without my consciously thinking about it. But then a chance encounter at a party reconnected the link.

I was milling around the patio at a party—trying to look comfortable in the company of thirty or so strangers—when I bumped into a familiar face from my hometown, an acquaintance really. For unexplained reasons he remembered what had transpired that week in 1973 when we found the plane crash and he eagerly asked me to "Tell the plane story." I resisted, telling him that it wasn't *that* exciting. He insisted. Soon, the pull of the story had captured most of the party goers and they circled around and listened intently. When I concluded the story, several people said that I should "Write this up sometime." Then, I caught a glimpse of a woman toward the back of the circle of people. She dabbed tears from her eyes and then she made her way over to me. Our conversation went something like this:

"Kind of a sad story," I commented, anticipating that was the reason for her tears.

She laughed while the tears continued to flow. "No. It is a beautiful story. Thank you so much for sharing it." I was confused at this point. Why had she cried? "You may not realize it," she continued. "But I think there was divine intervention at work here. This gives me hope." She smiled and thanked me again. It was clear that what I knew about the plane crash was powerful and it touched people.

Eventually, I decided I had to write an article about my experiences. But first, I needed to contact the families who lost men in the crash and let them know about this idea. Plus, there were a few details of the story that I had heard second hand and I wanted to verify before writing an article. It took awhile, but I tracked down Sara in Texas and sent a letter that asked if I could visit. Sara was my main hope for contacting the men's families after all that time: I knew that Sara had remarried and I hoped to locate her though him.

I pulled up to the house and the GPS in the rental car proudly announced, "You have arrived at your destination." For a town of only 4500 residents, this modern helper was hardly necessary but the GPS did lessen some of the anxiety I was feeling. A curved driveway led from the street and I followed. A short, trim brunette was sweeping the porch and I recognized her. Apart from some gentle aging, Sara looked the same.

"Sara?"

She looked up and smiled. "Bruce! It has been so long! Please, come in. Jack's inside." Sara married Jack after finally accepting that her husband Ken was not coming back, some three years after the plane's disappearance.

I walked in to greet a beaming Jack and a vigorous handshake. "It is so good to see you. You won't believe what we've got for you," he said. He eagerly led me into the den and pointed to a large hard-sided Samsonite suitcase on the floor. Wiry, and bursting with energy, Jack reminded me of a seven-year old eager to show his best friend what gifts he got for Christmas. "I don't know why we kept this stuff for all these years, but...." He popped the locks on the suitcase to reveal a thick stack of paper and artifacts from the plane crash.

What little anxiety I had about whether they would be willing to talk about the crash vanished in an instant. I sat down and began to comb through the pile of paper. I could hardly believe my eyes: Pages of original hand-written notes from 40 years ago had been saved. Unknown details about the search efforts flashed before me.

The scope of the search was huge, perhaps unprecedented for New Mexico. The wives played a large role in keeping the search active, undoubtedly with a significant emotional toll. It wasn't until my trip to Texas to visit Sara and Jack Rex when I realized there was so much more to the story, and the scope grew dramatically.

The story I'm about to tell you is entirely true. Much of the story comes from a reconstruction of events from hand-written notes contained in the Samsonite suitcase.

PART I

Mountains and Albuquerque

1

When Ansel Adams released the shutter on his camera in 1941 he captured one of his most popular and most sought after photograph in the field of fine-art photography: "Moonrise, Hernandez, New Mexico." It was a once-in-a-lifetime shot of a nearly-full moon rising over a mountain range that overlooks a small, quiet village.

There is a patch of white silken clouds hovering nearby in the sky, positioned seemingly to offer the moon a bed to rest if needed. The moon itself hangs starkly in the top of the photograph against a dark sky. The adobe church and white gravestones meld together in the gray landscape. The foreground and background each are strengthened by the other, illustrating the inseparable coexistence of mountains and valleys in northern New Mexico.

While the only signs of humanity are found in the valley, the mountains are never far removed from the peoples' lives. The snow-capped peaks remind the occasional visitor, however, that this can be a harsh environment. The vast scale of the landscape misleads the eye, and it is not apparent that the peaks stand over the valley by more than 7,000 feet.

The Rocky Mountains slice through the North American continent like a buzzsaw, abruptly rising from the relatively placid prairies of the Midwest, and interfering with the flow of people and weather. The Rockies stretch some 3,000 miles from northern British Columbia to New Mexico, terminating east of Santa Fe. Geographers refer to this southern-most segment of the Rockies as the Sangre de Cristo range; this is the range captured in the Moonrise photograph. It is the largest and highest mountain complex in New Mexico, with peaks over 13,000 feet in elevation and climatic zones comparable to Colorado. In his book *New Mexico Mountains*, author Robert Julyan notes that if you were to line up the tallest 71 mountains from New

Mexico, all of them in would be in the Sangre de Cristo range. The range is the state's longest, at 100 miles, not even counting that portion of the range that is found in Colorado.

Much of this story is centered within the Pecos Wilderness, one of the mountainous crown jewels near Santa Fe. The Pecos Wilderness is situated in the Sangre de Cristo range, extending east and north from the Santa Fe Ski Basin roughly in a square that is 25 miles on a side. The upper Pecos high country is named for the Pecos River, one of the great rivers of New Mexico, which starts in this region. Ponderosa pine, dense mixed conifer forests, small meadows, and rugged rocky terrain characterize the area.

Since prehistoric times, humans have visited this wilderness. However, their visits were only temporary, as the harsh climate of the ramparts forced them to retreat to the valleys during the winter. Today, use of the Pecos Wilderness is mainly limited to backpacking, hunting, and grazing. Apart from the meadows and alpine lakes popular with hikers, most of the Pecos Wilderness probably has not been touched by humans in the last century.

Although visited infrequently, these mountains have always had a spiritual and mystical power to people in the area. The Tewa Indians consider Lake Peak as the "sacred mountain of the East." They hold summer ceremonials at the lake on top of the Peak. Blue Lake, located at the southeast base of Wheeler Peak is sacred to the people of Taos Pueblo, who conduct ceremonies there not to be witnessed by outsiders. Holy Ghost creek empties from Spirit Lake into the Pecos River. The creek is said to have been named by a priest who was concealed from Pecos Indians by the mist that collects there, according to scholar T.M. Pearce.

General Search Area

Santa Fe "Baldy" Peak

Lake Peak near Santa Fe Ski Basin

2

The mid-1960s was an exhilarating time for technology. It had been only five years since President Kennedy had issued his challenge to the country : "I believe that this nation should commit itself to achieving the goal, before this decade is out, of landing a man on the moon and returning him safely to earth," and NASA was responding. The computer age was in its infancy, moving from research labs into colleges and industry. And the complexities of organ transplantation were being tackled by doctors worldwide.

It was ironic that the pursuits of risky space travel and the Cold War brought stable jobs to families across the nation. Increasingly, many of those jobs were coming to Albuquerque, especially for those related to technology.

It was the lure of working on cutting research in aerospace technology and defense-related problems, plus being in a place with great outdoor recreation possibilities, that brought three young engineers and their families to Albuquerque in 1966 and 1967. It promised to be the home for many years for the Brittain, Horton, and Jones families.

Gulton Industries was one of the nation's first conglomerates, with a scattering of 10 to 12 divisions performing a smorgasbord of diverse and sometimes seemingly unrelated tasks. One division specialized in instruments that would measure stress pressures in buildings or along earthquake zones. Another division made wire, while another built huge transformers for nuclear reactors or for experimental fusion reactors, and so forth. The Data Systems division in Albuquerque specialized in telemetry. Telemetry systems determine the health of the inner workings of a rocket, missile, or satellite. Because its capabilities were so unique within the company, the Albuquerque office largely was responsible for its own successes or failures.

The office began as a few-person operation but it quickly grew to more than 100 strong. While small compared to the major aerospace players, Gulton pioneered techniques to communicate with aerospace vehicles using state of the art digital technology rather than the prevailing analog technology. In addition to such computer systems breakthroughs, the company also became well known for their ability to manufacture solid state electronics components. As word of Gulton's successes spread within the industry, the company was able to compete for some of the brightest young engineers in the country.

Most of the young engineers arriving were in their twenties. In the developing fields of digital communications and computer science, the experts were young. Vice President of Gulton Data Systems, Ed Whaley was one of the old guys and he was only in his thirties. Before moving to Gulton, Whaley got his start in this field working at Bell Labs and at Sandia Corporation. Somewhat resembling a healthy version of actor Nick Nolte, his voice carried a soft southern accent from his Alabama upbringing.

Whaley was born to be a successful manager. He possessed the unique combination of understanding the complex technical elements of the business as well as being a good manager of his staff, helping them around any obstacles in the way of completing projects on time. Not only did he have a great memory, he took meticulous notes and always seemed to be on top of things. Under his tutelage, they became a premier telemetry systems designer for clients such as NASA, the Department of Defense, TRW, and Bendix.

Whaley only asked his staff to do their best for the good of the company, and they responded. He had a dedicated staff that was young enough to think they were bullet proof, voluntarily working late at night and over weekends if needed to finish a project. He expected a lot of his staff, yet they knew that he and the company would always be there for them in times of crisis.

Ken Brittain was one of the promising young recruits who joined the Gulton staff in the late 1960s. He came to the company after gaining valuable aerospace industry experience from working three years as an electrical engineer for NASA in Huntsville, Alabama and one year for Space Craft in Houston. Ken loved working in aerospace technologies and his career had blossomed in Huntsville and Houston. But he was a Westerner by heart. Soon, he came to miss the mountains and open spaces, along with the backpacking, camping, and fishing opportunities he had as a kid growing up in Arizona.

The answer to his dilemma came when he was offered a position as a senior project engineer at Gulton Industries in Albuquerque. He was excited to work on cutting edge digital electronics, particularly in their application to aerospace. So he readily accepted the offer. Ken, his wife Sara, and their young son Todd moved to Albuquerque in the spring of 1967.

It was during his junior year as an electrical engineering student at Arizona State University when he met Sara. To get romantically involved was the last thing she wanted at the time.

Unlike most of the rest of us, Sara knew precisely what she wanted to do after high school. Although raised in a small, rural town in Illinois, she had global plans.

Sara was shy by nature and a good student. Her real love was music. She enjoyed playing the piano for school performances, and she played the flute in the band. Sara stood just a few inches over five feet. Her dark sweeping eyebrows complemented an easy smile. However, her physical presence is not what got your attention. Instead, it seemed to be the gentle grace that she carried herself with.

For many kids from the community, continuing on to college was a major undertaking. So it certainly must have come as a surprise to some when she told her friends and family that she planned to go to college and study music, and then become an International Missionary for the Baptist Church. Coming from such a stable, rural, heartland American town where only about one in 10 adults held college degrees, this was a bold plan.

And so it was with this lofty goal of being a foreign missionary that she enrolled at Grand Canyon College, a private Baptist-founded school in west Phoenix, Arizona. Sara would study Music Education.

It was early during her junior year in college that her long-term plan began to be challenged. One evening Sara was to receive an award at a campus-wide gathering. In attendance at the gathering was a young man who took notice of Sara as she received her award. Ken Brittain was a studious junior engineering student at Arizona State University, located on the opposite side of Phoenix, and was visiting her campus that evening. Shortly thereafter, Sara received a phone call in her dormitory from Ken. When he introduced himself on the phone, she realized that his sister Babs attended Grand Canyon College. After much persuading, Ken finally got Sara to agree to go on a date "to do something, maybe go to church."

From the first, Sara found him very comfortable to be around. They dated through the year, and things got serious. The summer following their junior year, Ken proposed. But Sara was going to be a foreign missionary. "So I just didn't know what to do about this," she recalled. Ken persisted, and later that summer he drove to Illinois to meet her family. He proposed again there and was prepared for a third try if she hadn't accepted. They became engaged at the end of the summer. After graduation they married and moved to Huntsville. Sara quickly began to use her music skills. She taught music in the Huntsville Public School system and played music for church services. However, with the delivery of their first child, Todd, Sara stopped teaching in the schools and became a full time mom. Sara's plan to become a Missionary was derailed—or at least postponed—by romance. With Ken she had found her life partner.

The State of Texas is huge, blessed with diversity in vegetation, landscape, and accents. However, there are several characteristics that Texans often share. There is a high degree of self-confidence ("Things are Bigger in Texas," "Don't mess with Texas"). There is hard work. And there is a zest for adventure outside of work. Visit a seemingly out-of-the-way spot in the mountains of Colorado and you shouldn't be surprised to come across a group of Texans looking for fun. Folks think nothing of leaving Dallas on a Friday afternoon, driving all night to get to an alpine ski area, only to return in time for work on Monday morning.

To a great extent, these characteristics were seen in Beverly and Jon Dale Horton, both native Texans.

She first saw him at Arlington State College (later University of Texas at Arlington), a good-sized public two-year school located between Dallas and Fort Worth. Beverly was studying commercial art and Jon Dale was an engineering student. Arlington State had started as an all-boys school under the A&M system that emphasized engineering, and the male to female enrollment "was about ten to one for me," recalled Beverly.

Despite the large number of male students to compete for her interests, Beverly singled out Jon Dale the first time she saw him while she was working a desk at the student center. He was tall, well groomed, and wore a stylish suede jacket. Over time, they began to date.

Both Beverly and Jon Dale were athletic and competitive. At nearly 5'8", Beverly was long and muscular, while he stood 6'3" and had a bit of a tough cowboy look. She was a member of Arlington State's tennis team and he had played High School football. On one of their first dates, he agreed to play her in tennis, probably thinking that his athleticism would make up for his lack of skills in tennis. She crushed him. For Jon Dale, this couldn't stand. So he began to quietly practice tennis with other girls on campus, in anticipation of a rematch. This scheme came to a quick stop when Beverly discovered his plan.

While engineering is a demanding major, Jon Dale tried to add adventure to his school time. One weekend he and some friends left Arlington and drove to a lake to do some hunting. However, because of the school load, they brought their books along for the 'quiet times'. After they arrived at the campsite along the lake, they put up their tent and built a fire—a huge fire that burnt down the tent. They lost all of their books in the fire, along with the famous suede jacket. No worries. They made it through the weekend and finished the school term without having to buy new books.

For Jon Dale, this was the start of many years of straight schooling. Next stop for him was the University of Texas at Austin where he received Bachelor's and Master's degrees in Electrical Engineering. Through this time, they dated on a start and stop basis, and occasionally talked about marriage. Although each felt they loved the other, the physical separation of him being in Austin and her in Arlington was a major complication in the relationship.

Jon Dale left Austin after he finished his Master's and drove to his first major job in Las Vegas, Nevada. He took a position with EG&G (at the time called Edgerton, Germeshausen, and Grier, Inc.), a high technology consulting firm headquartered in Boston, Mass. that supported defense related research activities. EG&G was legendary in the electronics field, having pioneered such breakthroughs as the strobe light and side-scan sonar.

To celebrate his graduation and new job, Jon Dale gave himself a big blue Starcraft, inboard/outboard 18½ foot ski boat. He named the boat "The Beverly II," for the girl he was in love with, he said. (Beverly I was a small green canvas kayak that he and his friends each built for themselves over spring break.) He and his buddies would ski after work in the summer months and on weekends. When Beverly visited, they would take the boat out and camp on the lake on the weekends; Sunday night, Beverly would take a red eye flight home in time for work on Monday morning.

Although he lived in Las Vegas, Jon Dale's actual work site was 80 miles away at the Air Force's most secret test facility at a remote piece of

Edwards Air Force Base, Groom Lake, more commonly known as Area 51. According to one of his co-workers, he was assigned to carry out electronic measurements to measure the radar cross section on prototypes of the SR-71 spy plane, one of the nation's highest classified projects.

In little time after moving to Las Vegas, he greatly missed Beverly and he thought more about marriage. Nonetheless, he explained to Beverly that he decided against marriage for now because of the flying risks of his job. Early in the morning he and other scientists would board a plane in Las Vegas and fly to remote sites. He worried about the maintenance on the planes. In a prophetic comment, he told Beverly that "Flying to the site (Area 51) is too dangerous and I wouldn't want to do that to you."

One bright spot of his work at EG&G was the friendships he and Beverly developed with his co-workers, particularly the Rieffs, the Jones, and the Sundines families, who later came to the Albuquerque area and either worked for Gulton Industries or EG&G. Amongst those co-workers were fellow electrical engineers Ron Jones and Dick Reiff. With Jones, he discovered a shared interest in the outdoors, particularly in hunting. When possible, on weekends they hunted the varied terrain of southern Nevada.

After one year, testing on the SR-71 had finished and Jon Dale began to look for work elsewhere. When his Las Vegas friend Dick Reiff landed a job with Gulton Industries in Albuquerque, he immediately recommended Jon Dale for recruitment. Jon Dale jumped at the offer from Gulton. Not only would he be closer to Texas there, but he and Beverly could finally go ahead and get married, after a seven year courtship. They married in Dallas, Thanksgiving weekend, 1966. Next stop after the honeymoon—a snow skiing trip to Vail and Aspen in Colorado—was Gulton Industries in Albuquerque.

As it coincidentally turned out, Jon Dale's hunting friend, Ron Jones, also moved to Albuquerque about the same time. Jones did not change employers, but remained with EG&G and transferred to the Albuquerque office.

Ron Jones was a classic Florida kid. Handsome and athletic, he spent every available moment outdoors playing on the fields and in the water. He was raised on Florida's west coast in Bradenton, situated at the mouth of Tampa Bay. He probably got much of his love of nature from his dad. Although his childhood home sat just minutes from the ocean, his dad often would rent a cabin on a nearby island for holidays, simply to share being outdoors with his family. Fortunately for Ron, the importance of education also was thoroughly stressed to him, particularly by his mother who was an officer in the local bank.

Armed with an electrical engineering degree from the University of Florida, he signed on with NASA as a junior engineer. As a part of his job, Ron Jones frequently visited Pan American Airway's 'Tech Lab.' It was during one of his visits that he met his future wife Barbara, who was doing secretarial work at the site. She was raised in the town of Cocoa on the east coast of Florida, just a few miles from the gates of Cape Canaveral. Working at Pan Am allowed her to stay close to her parents, and continue her education at nearby Brevard Community College.

Soon the couple was blessed with their first child, Ronnie, and Barbara became a full time mother. Whether for professional or personal reasons, Ron decided to look beyond NASA for employment. They decided to leave Florida and relocate to Las Vegas, Nevada where Ron accepted a position with EG&G, supporting the SR-71 project at Area 51. Their second son, Richie, was born in Las Vegas.

The landscape and culture of Las Vegas was far removed from what they were used to in Florida. Being away from parents was difficult for them and they returned home when possible. At the same time, living in southern Nevada gave them opportunities. With two young kids now, Barbara threw herself into raising the boys and helping out at the Baptist Church. She was busy and her life was fulfilling.

Even though they lived in the 'entertainment capitol of the world,' Ron was enthralled with the deserts, mountains, and rivers near the city. With a growing network of friends from work and church, Ron was increasingly asked to join them on short hunting trips. Hunting became one of his prime passions. His first love, however, was always family; wherever Ron's work would take them, they always would have each other and their families to unite them. The next chapter in their lives took them to Albuquerque when Ron transferred EG&G offices.

PART II

Gone Missing

3

September, 1968

Seven Bar Flying Service was about the only game in town for pilots of small aircraft. Located at the Alameda Airport on Albuquerque's expansive west side, Seven Bar was a full-service facility that offered pilot flying lessons, rentals, and maintenance for small aircraft. By today's standards, the airport was small, with one main building, a couple of hangars, and a small airstrip. Nonetheless, it was ideally situated for student pilot training. Visibility was often greater than 50 miles and it was near the western edge of the city away from the heavy commercial aircraft fly ways.

Many small aircraft pilots in the Albuquerque area took their flying lessons at Seven Bar. The facility allowed for budding pilots to not just obtain top notch flying instruction, but one could also rent planes to practice and hone skills. Typically, the student would train in a two-seater aircraft and then graduate to larger aircraft after they received their basic pilot license.

One recent graduate of Seven Bar's flight school was 27-year old John Warren Fishel. He had been bitten by the flying bug about a year ago, and was working hard on his flying skills. While airplanes were his current interest, Fishel loved machines of any kind. Before airplanes, it was cars. This passion came from his father who was a tractor and automobile mechanic in rural eastern Nebraska. As an adult, John always seemed to be driving a different car each time he visited the family, his oldest sister Betty recalled; his last car was a Pontiac GTO with a custom paint job. John Warren was smart, but it seemed that he was more destined to work with his skilled hands after high school than to pursue academic studies. Six months after his high school graduation, John Warren enlisted in the Air Force and served three years in Vietnam. His love of flying probably developed when he served in the Air Force.

After his discharge from the military, he went to work for EG&G. Fishel's innate mechanical aptitude and military training in electronics made him valuable to EG&G. He met and married a girl from Boston during that time.

His apparent good fortune was short lived. Tragedy hit twice. The couple had two baby boys but each only lived a short while after birth due to undeveloped lungs. The couple divorced only to remarry and divorce again as they realized that their marriage was truly over.

Undoubtedly devastated by his home life in Boston and desiring a new start, Fishel transferred to EG&G's Forth Worth, Texas office, and soon to the rapidly expanding EG&G offices in Albuquerque. He embarked on a fresh life. The EG&G technician work provided professional long-term stability and challenges to him. Not one to sit idle, outside of work he took advantage of the recreation possibilities around Albuquerque. The first interest that he developed was alpine snow skiing, until he broke his leg during his inaugural season. Next came flying.

Joe Hoffman, a fellow technician, recalls that John enjoyed talking to other military veterans in the EG&G offices. Like Fishel, they were entitled to post-military training benefits through the G.I. Bill. It intrigued John that many of these veterans had chosen to use the GI benefit to obtain their private pilot license, rather than for traditional college courses. Soon he made the decision to do the same. He wanted to fly and maybe even become a commercial pilot someday.

John threw himself into flying lessons. Fishel often could be found at Seven Bar squeezing in some flying time before or after work, and on weekends. In short order he had successfully completed the written and hands-on requirements for the private pilot license. While most of his flying was done around the Albuquerque valley, he occasionally ventured beyond the familiar. One time he came to North Bend, Nebraska and flew his parents to Davenport, Iowa, where his sister Betty lived. His confidence as a pilot was growing but he was cautious. At one time, John's mother said he felt comfortable behind the wheel but the mountains scared him.

Now that Fishel had obtained his private pilot license, he would have to bear more of the costs of renting planes as he gained flying hours. It was particularly expensive to rent one of the larger planes, those capable of carrying four people (the next step up).

One September day, after listening to his co-workers talk of the upcoming hunting seasons, he came up with an idea that might allow him to gain some flying time in one of the four seaters. He would recruit three other hunters to join him and they would fly into the mountains to look for possible areas to hunt for elk. He had two weeks to recruit his passengers before the elk hunting season opened. The advantages of his idea were clear. The costs would be covered mostly by the passengers, he would gain some valuable flying time in the mountains, and the hunters would increase their odds of success by spotting the locations of elk herds from the air, in advance of the hunt. A win-win situation for all.

One of the first asked to join the trip was Jack Frost, a fellow EG&G worker and hunter extraordinaire. Frost was scheduled to be working in Hawaii during the elk season and he reluctantly turned down the offer. However, he offered John Fishel some recommendations on where to hunt for elk, if Fishel himself was going to hunt. Frost recommended that they hunt Hamilton Mesa and Round Mountain, located within the southern portion of the Pecos Wilderness. Situated at altitudes greater than 11,000 feet, the area contained beautiful aspen groves and large meadows. Hunters could access the wilderness boundary by driving the 10 miles from the village of Pecos, passing Dalton Canyon and Holy Ghost Canyon as they climbed into the mountains.

Fishel continued his recruiting of EG&G people. It was slow going. His fellow technician Joe Hoffman wasn't planning to hunt that year, while others had family commitments.

Barely a week away from opening day, he got his first passenger. Ron Jones, John Fishel's supervisor at EG&G, agreed to join in the flight. Jones originally wasn't planning to go elk hunting this year until his interest surged after he bagged an antelope over the weekend. Fishel told Jones that the flight still wouldn't happen unless they filled all four seats in the plane.

Gulton Industries employees Jon Dale Horton and Ken Brittain had decided more than a month before they would join forces and hunt for elk. Being engineers, typically a meticulous breed, they planned the trip thoroughly. They took advantage of coffee breaks at Gulton Industries to talk about the hunt or to pulse other Gulton workers about where to hunt. After considering a half-dozen possible hunting locations, they narrowed their choices by taking long hikes on the weekends to find the ideal spot. Horton created detailed hand-drawn maps which would help guide them back to promising spots. Eventually they agreed to try the mountains outside of a small village named Tres Ritos. The village sits within the Carson National Forest, roughly mid-way between Santa Fe and Taos, New Mexico. The General Bull Elk season would run in this area October 5 through October 13.

The hunting area was well chosen, both for terrain and for access. Tres Ritos is located within the spine of the Sangre de Cristo mountain range, and only five miles north of the Pecos Wilderness boundary. Tres Ritos is situated along State Route 518, one of the few paved roads that traverse the mountain range in an east-west direction, ensuring year-round vehicle access to the remote area. The overlooking mountains contained a mosaic of mixed grassy meadows and forest, ideal habitat for elk which are considered to be "forest-edge" animals.

For the engineers, weekend camping or skiing trips with their spouses were common and camping often was the center of family vacations. While the wives normally did not accompany their husbands on hunting trips, the elk hunt this coming weekend would be an exception. Beverly Horton and Sara Brittain would relax at a rented cabin in Tres Ritos while "the boys" hunted during daylight. At least the wives hoped they could relax—it was getting more difficult as their families grew. Sara would be joined by her two high-energy boys, Todd and Billy, ages three and one. And Beverly was three months pregnant with their first child.

It was Monday, September 30, and all was set for the coming weekend hunt for Ken Brittain and Jon Dale Horton. However, when Jon Dale got a phone call from his old friend Ron Jones, an exciting opportunity was presented. Ron turned to Jon Dale to invite him on the flight because of the friendship they developed when working together at EG&G Las Vegas and at Groom Lake.

Excitement poured from Jones' voice on the phone. He first told Jon Dale about his antelope hunt the past weekend in southeast New Mexico. Within an hour after sunrise on the opening day of the hunt he was successful! Now he was anxious to shift gears and try for an elk. Originally he wasn't planning on hunting for elk, but the antelope hunt energized him and changed his mind. While Ron wouldn't be joining Ken and Jon Dale on their hunt in Tres Ritos, he had an idea that could help all of the men.

Jones said he had a friend "who is about to be a commercial pilot and will take me up in a plane to spot for elk," but he needed four people in the plane. With little hesitation, Jon Dale thought this was a great idea and he agreed. That he would consider such a last-minute twist was a testament to the friendship he had with Jones. Indeed, it was their professional lives that provided the common threads that pulled their social lives together.

Jon Dale had read about flight hazards, but he was comfortable with flying as he knew it. Some of his closest school friends in Texas became pilots: one ended up flying for Braniff Airlines and another for the U.S. Marines. His friends used to joke about the risks with Jon Dale and Beverly: "Do we go over or under the wires?" Beverly had dated some pilots in the past and had even seen one ditch a plane in a pasture during an emergency. Still, their experiences were in the flatter parts of Nevada and Texas mostly, and they didn't know much about mountain flying.

That evening, Jon Dale told Beverly about the possible flight and that a fourth person was needed to join the party. Beverly volunteered right away to be the fourth person. "It sounded like a fun thing to do," she recalled. However, she had to check her class schedule first. And, then there was her pregnancy: Jon wondered if it was wise for her to go because she was expecting. Jon's response left Beverly conflicted. She wanted to go on the flight because it might be fun. On the other hand, maybe he did have a point about the pregnancy.

Leaving the topic unresolved, Beverly said she would think about it. Just like all young couples, they were being faced with changes to their life styles as their families grew. Perhaps arising from this conflict, she told Jon in the ensuing conversation that he "...would have to stop going on these adventures" and he responded "Don't lay that on me." By the time Beverly decided to go on the flight, Jon Dale said "Don't worry about it. I got Ken (Brittain) to go." They would depart Wednesday morning, October 2nd.

4

WEDNESDAY, OCTOBER 2ND, 1968

Ken Brittain awoke at four a.m. and quietly got dressed so as to not wake Sara. The men were expected at the airport in less than two hours. He wore a light windbreaker over clothes that were more suitable for the office than for the mountains to which they were headed. After they got back from the short flight he would drive directly to the office for an important pricing meeting at Gulton. As he got into his car, the nearly full moon had just set on the west side of the Rio Grande Valley and the early morning was crisp with only a light breeze. He had three passengers to collect before the flight. Fortunately, all the men lived in the Northeast Heights of Albuquerque and the collection was easy. The first passenger was Jon Dale Horton, and the two then drove to Ron Jones' house for some coffee.

Eventually, the friends drove to pick up John Fishel, the pilot. Only Jones knew John Fishel and the others relied on Jones' endorsement before deciding to make the trip. They would be leaving out of Seven Bar Flying Services and Alameda Airport on the western edge of the city.

From the Northeast Heights, the men drive down the gentle slopes towards the Rio Grande. As they approached the river they passed through the world's largest stand of cottonwood trees, called the Bosque. The heart of Albuquerque's downtown lies along the river, but the urban effects were sheltered by the Bosque. In a few minutes more, the men had arrived at Seven Bar.

John Fishel had recently received his private pilot license through the instructors at Seven Bar. He continued to log more hours in hopes of eventually obtaining a commercial license. Records show that he had logged 115 hours of total flying time by now, more than double the minimum 40 hours required for a private pilot license. However, the majority of this time was logged flying around the relatively flat Rio Grande Valley. As much as Fishel loved flying, it still was a part-time activity until he gained more experience.

Today, John Fischel rented an airplane from Seven Bar for the early morning flight, just as he had on several other occasions. He told the facility that he was doing training. It was a Piper Cherokee 180-hp single engine airplane of white color with blue trim.

At three minutes after six a.m. the pilot telephoned the Albuquerque Flight Service Station to file a Visual Flight Rules, or VFR, flight plan and also to get the reported weather in the Santa Fe area. According to VFR restrictions, the weather conditions must be favorable and allow the pilot to visually navigate by looking at the environment outside of the cockpit, control the altitude of flying, and avoid any obstacles.

John Fishel was advised that the weather at Santa Fe was good with just scattered cirrus clouds and all surface winds below 10 knots. Winds were from the northwest at 25 knots at 8000 feet, and were 20 knots at 10,000 feet. In what may have been the most important footnote in the weather report, there was a caution of strong downdrafts on southeast facing slopes of the mountains.

After listening to the weather forecast, John Fishel decided to proceed and he verbally filed a flight plan for the elk scouting expedition. The flight plan relayed in the telephone call contained the following information:

True Air Speed: 110 mph
Departing Alameda: Estimated at 6:15 a.m.
Route: Pecos Mountains northeast of Santa Fe
Time Enroute: 2 hours
Fuel on Board: 3½ hours

The men loaded into the Piper Cherokee; two in the front seats and two in the rear. Fishel radioed to activate the flight plan and the aircraft departed at six forty a.m. Return to the Alameda Airport was expected by eight forty a.m. Another small plane that departed immediately after the hunters reported that the Piper Cherokee flew east and then north, in the general direction of Santa Fe.

ALBUQUERQUE, NEW MEXICO, UNIVERSITY OF NEW MEXICO CAMPUS,
OCTOBER 2ND, 1968

Beverly Horton was feeling good about her life. After a roller coaster seven-year courtship with Jon Dale, the couple finally married 16 months ago. They seemed to be on the move nearly every weekend; skiing, boating, or hunting. He was trying his hand in spelunking and rappelling. Most important of all, she was three months pregnant with their first child. Blessedly, it had been an easy pregnancy so far. There had been no symptoms of the dreaded morning sickness.

She also was excited about this coming weekend. Soon she would be relaxing in a cozy cabin at Tres Ritos with her friend Sara Brittain, while their husbands did their elk hunting thing.

But first she had to do get through a few days of classes at the University of New Mexico. Beverly had the creative eye in the family, and she was studying Art. She arrived on campus before nine a.m., got coffee and headed to the first of three classes for the day: Art, English, and Anthropology. The Art class was unremarkable.

Beverly arrived at her second class to find the classroom still full of students. As they waited for the room to empty, Beverly chatted with one of her friends. Suddenly, with no obvious explanation, a wave of dizziness hit her. It was unlike anything she had experienced before and she nearly collapsed to the floor.

Lightheaded, Beverly Horton was steadied by a classmate who helped her to a chair. What caused the sudden blast of weakness? Immediately Beverly began to think that this might have been caused by morning sickness, yet throughout her pregnancy she did not experience any symptoms like this. Indeed, in the subsequent decades, Beverly never again experienced such a sensation.

As manager of the Seven Bar Flying Service, Robert Jannson had barely finished his first cup of coffee when they came to his door. He could tell when his staff had a routine question. This time was different.

It was nine ten in the morning when Seven Bar personnel noticed that rental N 8905J was overdue, and preliminary search procedures were implemented. The airplane was 30 minutes overdue on flight plan. Repeated radio calls to the aircraft yielded no responses. A walk around the grounds of the airport showed no signs of the missing aircraft. Efforts to contact the aircraft continued. So far, silence.

Jannson wanted to know everything about the pilot and the flight plan. His staff was ahead of him, and they handed him a sheet of paper. The flight plan was studied to better understand the pilot's intents. The plan that John Fishel had phoned to the Albuquerque Flight Services Station (FSS) frustratingly lacked detail, stating only "Pecos Mountains northeast of Santa Fe."

Shortly thereafter, the pilot's intent became clearer after one of John Fishel's flight instructors phoned in some more detail. Nick Beers remembered chatting with Fishel about a hunting trip that would occur in a few days. Although Fishel did not mention it when he rented the airplane, it became evident that the plane was used to help the hunting effort. Jannson and Beers both knew that Seven Bar didn't authorize flights for hunting purposes.

Ed Whaley arrived for work at Gulton Data Systems around eight o'clock. His calendar for October 2 was fully booked. The most important item on his list this day was to complete negotiations on a proposed project with a company from the St. Louis, Missouri area. Preliminary cost estimates for the project had already been given to the company and today Gulton Industries would follow with a detailed budget quote that Ken Brittain had worked up. Ed Whaley and Ken Brittain were to meet at eleven to review the quote and ensure that all details were addressed before phoning the client in the afternoon.

Whaley went directly to Ken Brittain to make sure all was set for their review meeting later in the morning. But, Ken wasn't there, so Whaley checked around to see if he was sick or had called in. Ken's secretary told him Ken had left word that would be in for the eleven o'clock meeting, and that he was going on an airplane to look for a place to hunt for elk, but he'd be back in time. Ed Whaley was busy the rest of the morning, but around ten thirty he went by to check if Ken Brittain was in, because he was getting a little concerned. "So, when he wasn't back by eleven a couple of us got together the data (Ken's budget workup) that we had and started to look it over. We decided we were comfortable going ahead with the bid, although it would have been better if Ken was there."

Beverly Horton recovered well enough from her dizzy spell to sit through her second class. Then with one class remaining, Beverly headed to her car to try her luck with eating lunch. As was her tradition, she drove to a drive-in restaurant near campus to eat a sandwich.

If nothing else, she thought, at least she could listen to the radio in the car. Coverage was just starting of the opening day of the World Series between the St. Louis Cardinals and the Detroit Tigers. In a highly anticipated match up, the two top pitching aces in the game were going against each other; Bob Gibson of the Cardinals and Denny McClain of the Tigers together had combined for a whopping 53 wins and 19 shutouts during the regular season.

Deciding that she'd rather listen to some music or hear some news, she reached to the radio and changed channels. In a breaking news bulletin, the announcer told his audience that a small airplane was missing and search efforts would soon start. Beverly felt chilled. With uneasiness she wondered if it was Jon Dale's plane. On the way back to work from lunch, Ed Whaley heard the same missing plane radio broadcast as Beverly.

Beverly tried not to panic. She recalled an incident early in the year when Jon Dale was late to return from grouse hunting. When Jon Dale eventually walked in the house, she asked "Why didn't you call?" He simply said, "Well, we had a flat tire." Beverly knew he didn't like to be checked up on. She wrestled with whether she should race home to see if Jon Dale's car was there, or to relax and return to class. If this was *not* Jon's plane, she could just hear him say "Well you drove all the way home and you missed class." So she went to her next class, but she "didn't hear a word" because of worry. Adding to her frustration, she didn't know which airport they were taking off from, so she could call and check on them.

Beverly drove up to her home between two and three in the afternoon, and Jon Dale's blue van was there. Her heart dropped. She knew that he would've driven the van to work after the flight, and this probably meant that it was his plane that was missing. Desperately hoping for some good news, she called Gulton Industries and asked to speak to Jon Dale. Ed Whaley came on the line and said that neither Jon Dale nor Ken Brittain was there. Beverly explained to Whaley what she had heard on the radio. He asked her if she knew the name of the pilot. She did not, but she would call Barbara Jones and find out.

It was Ron Jones' birthday, so Barbara had been out all day running around getting things together for the party and tending to her two little ones. Barbara was surprised at the call, but she did know the pilot's name along with the airport they planned to fly from. After relaying this information to Ed Whaley, Beverly then phoned Sara Brittain.

"Sara do you know where the boys are?"

Sara was busy with her two kids and hadn't given it any thought. No, she hadn't heard anything.

At one thirty-six p.m., the U.S. Air Rescue Center at Richard-Gebaur Air Force Base in Grandview, Missouri was notified that the single-engine Piper Cherokee N 8905J still was not located and suggested that search and rescue action be initiated. The Rescue Center was the official office through which all Civil Air Patrol Search and Rescue missions across the nation were coordinated. This included New Mexico. Immediately the Air Force Rescue Center notified two groups about the need to initiate a search. First, a call was placed to the Civil Air Patrol (CAP) New Mexico wing, officially activating them. The CAP is an all-volunteer organization overseen by the U.S. Air Force. The CAP members volunteer their time, talent, and privately-owned equipment for missions and are compensated only with gas and oil. Next, the New Mexico State Police were alerted, triggering emergency protocols for other state and federal agencies.

Now that Ed Whaley had the names of the pilot and airport, his first stop at Gulton Industries was his old friend Larry Neely. They talked about whom to call to see if it was 'our plane' that the radio broadcast referred to. As Ed Whaley recalled:

"The authorities wouldn't give us the names of anybody on the missing plane. Larry said, 'I think I can get around that.' He had a very close, personal friend who was high up in the State Police. So Larry called his friend, and said 'Here are the people from our plant that were on a plane and failed to return. Can you say if these people are on that missing airplane?' And the friend said 'Yes, they are on the airplane.' By providing the names and asking the police to confirm them, Larry didn't put his friend in a position whereby he would be violating police procedure. We gave them the names and the police confirmed. The friend said, 'I can't say anything more, but they are on the plane we are looking for.'

For the people at Gulton Industries, there was never any doubt they would help find their colleagues. As Ed Whaley put it, "They were our people."

Beverly waited until the evening to call Jon Dale's parents, her in-laws, at home. The last time they spoke to his parents was to announce that she was pregnant, so the Hortons were expecting more good news. She told them she didn't really know much, but she hadn't heard from the men and the names were soon to be released. She didn't want them to hear the news from someone else. The Hortons responded that they were leaving for Albuquerque the next morning. Beverly asked them if they could bring her mom with them, which they did.

By dinner time, news coverage about the missing plane broadened. The evening TV news in Albuquerque had stories describing the four men in the plane and the newly started search efforts. The early afternoon newspaper, the *Albuquerque Tribune*, quickly inserted a six-paragraph story onto page 1:

Small Plane Overdue, Four listed Onboard

Four persons in a blue Piper Cherokee single-engine airplane—who left Albuquerque early today on an elk spotting trip to the Pecos area—were reported overdue this afternoon.

The pilot was identified by Robert Jansson of Seven Bar Flying Service at Alameda Airport as John W. Fishel, 27, of 720B Truman NE.

Greg Islas, 22, who shares an apartment with Mr. Fishel, said the pilot "got his license a couple of months ago."

Mr. Islas said Fishel is employed by EG&G and planned to take some fellow workers with him today to look over elk-hunting areas in preparation for the winter hunting season.

The Civil Air Patrol was to start a search for the missing plane this afternoon.

Mr. Jansson said Fishel—whom he described as a good pilot—left Alameda Airport about 6:30 a.m. for a three-hour flight. He had three hours of fuel on board.

Mr. Jansson said Seven Bar doesn't authorize flights for hunting purposes."

5

The three wives tried to maintain some calm that evening, despite the deep fog of uncertainty. Each dealt with the nightmare somewhat differently, but they shared hope that the men would all be fine. Beverly Horton and Barbara Jones stayed home and were supported and comforted by their closest friends. The CAP called with the latest search news and the wives, in turn, kept busy phoning updates to parents and relatives.

As soon as the names were released, Gayle Harris was at Beverly's house to stay the evening. Her husband, Ed Harris, had shared an office with Jon Dale when he first came to Gulton, and the two couples often danced or camped together.

Rather than stay at home like the other wives, Sara Brittain headed off to church for a few hours. This night was the last of three consecutive nights that she was to play piano at a church revival, and she felt an obligation to follow through with her commitment. Sara felt that going ahead and playing at the revival helped maintain some normalcy in her life. It also showed that she expected him to walk in the house, as if nothing had happened that day. After the revival, she returned home and was quickly surrounded by friends.

The following morning, Beverly was awakened by a ringing phone at five a.m. She "just knew it would be Jon telling me he was alright and would be coming home." Instead, it was a newspaper reporter, who wanted to be the first to get his information.

6

At this point, the wives were still adjusting to the *possibility* that their husbands were missing and the *possibility* that they could have been involved in a plane crash. Nonetheless, there was something about a missing airplane that unnerved them, more than if, say, their husbands were late in returning from an automobile trip.

Since the start of aviation, people have been conflicted about flying—both fascinated and frightened. After all, flying is the most unnatural of human activities. Beginning in 1903 with the Wright brothers' first successful flight, it had taken some fifty years before the public began to feel somewhat comfortable with the notion of flying as a means of traveling. The industry got a short-term boost from World War I advances, but government funds dried up after the war, and growth of the industry slowed. Interest in flying, however, never waned because of the occasional dazzling event like Charles Lindbergh's flight across the Atlantic Ocean, and the start of delivery of air mail by the U.S. Postal Service. The biggest boosts in aviation design came from World War II research and development efforts. The subsequent advent of four-engine aircraft and the jet engine greatly reduced flying times with improved safety, and the commercial airline industry was born.

Still, with all the advances, there remained deep-rooted concerns about safety. The occasional airplane crash continually reinforced the fear. The risks of flying were amplified when a celebrity or public figure was killed. Such was the case on February 3, 1959, when Buddy Holly, a rock 'n roll founding father, was killed in a small airplane crash in Iowa along with J.P. "The Big Bopper" Richardson, Ritchie Valens, and the pilot. After recording "Peggy Sue" in a small studio in Clovis, New Mexico, Holly had become an international star and his sound would revolutionize the world's music scene.

Amidst a whirlwind bus tour of twenty-four Midwestern towns and cities, the flight was taken because the bus was not equipped for the winter weather. The plane crashed shortly after taking off in the poor weather conditions. The pilot was not qualified on instruments, and was using Visual Flight Rules. The plane crash has been called the greatest tragedy that rock and roll has ever suffered. The day was called *The Day the Music Died* by Don McLean in his song "American Pie." These pioneering artists and the crash were further immortalized in the movie "La Bamba."

The 1960s brought a string of other high profile crashes. In 1960, the US Navy Band and 16 members of the Cal Poly San Luis Obispo football team were killed in separate crashes. In 1961, 18 members of the U.S. Figure Skating Team and the U.N. Secretary Dag Hammarskjold were killed in separate crashes. Singers Patsy Cline, Jim Reeves, and Otis Redding, golfer Tony Lema, and boxer Rocky Marciano died in later crashes. And the decade closed out when all 74 people traveling with the Bolivian Soccer Team were killed in 1969 upon hitting a mountain peak.

On the commercial aviation side, industry and the federal government worked steadily to lessen the rates of aviation accidents through improvements in technology, maintenance, and training, making commercial air travel amongst the safest form of mass transportation. But the story was very different in the world of general aviation, which included small private aircraft. Every year, many hundreds of people were killed in general aviation accidents, and thousands more were injured. According to National Transportation Safety Board (NTSB) studies, general aviation continued to have the highest aviation accident rates within civil aviation, about six times higher than small commuter and air taxi operations and over 40 times higher than larger transport categories.

Fatalities from general aviation crashes in the United States peaked in the period 1965 through 1980, according to a research branch of the U.S. Dept. of Transportation. Total yearly fatalities for the years 1965, 1970, 1975, and 1980 were each greater than 1000 souls. Since then, the number of fatalities declined progressively to 450 in 2010. For comparison, about 50,000 people died each year in motor vehicle accidents.

In New Mexico, plane crashes and accidents were common in the 1960s despite its relatively low population. Over the 10-year period 1964-1973 (bracketing the time of this story), there was an accident every month, according to NTSB records. Each year there was an average of 88 crashes and 20 fatalities. Of the fatal crashes, 85% involved aircraft with a single engine.

Given this constant string of accidents, along with the geographic isolation of most of the state, the New Mexico CAP had considerable activity in Search and Rescue. In 1965, New Mexico CAP ranked fifth nationally in the number or sorties and in total hours flown during the year. Most of the searches appeared to be relatively short, typically lasting one or two days. Among the 91 fatal crashes listed in the NTSB database for the 10-year period, four were flagged as lasting longer than one week and only two crashes took longer than three months to recover. Over this period, the New Mexico wing of the CAP flew an average of 330 sorties and 765 hours in search and rescue operations per year.

7

The Civil Air Patrol search efforts were coordinated out of an adobe office building situated by itself on the grounds of the Santa Fe, New Mexico Airport. Lieutenant Al Lopez was the CAP mission coordinator. When he was not involved in Search and Rescue operations, he worked as a chemist at the Los Alamos Scientific Laboratory.

Lopez knew that time was their enemy. Survival rates for crash victims rapidly diminished after just 48 hours. He was well aware of the grim statistics:

> Only 29% of souls on board a small aircraft will survive a crash, on average.
> Of those that survive the crash, 60% will be injured.
> 81% of the injured will die if not located within 24 hours and 94% will die if not located within 48 hours.
> Half of the uninjured survivors will die if not located within 72 hours.

The team of Al Lopez was to designate search areas and patterns that had the highest probability of locating the missing plane, safely and quickly. The starting point for the CAP planners was the flight plan phoned to the Albuquerque Flight Service Station by John Fishel. The brief flight plan provided few concrete details, which could reduce the effectiveness of the search. Nick Beers, one of Fishel's flight instructors, told the Flight Service Station that he thought a more accurate description of the intended route was: Alameda Airport to Truchas Peak to Cerro Vista Peak and, time permitting, Tres Piedras.

The sheer size of the search area quickly became apparent. From their departure in Albuquerque, the men could have traveled nearly 200 miles to the north near the Colorado border and could have possibly returned along a line some 70 miles farther to the west. As an initial estimate, approximately 13,000 square miles of high-altitude mountainous terrain would need to be searched, an area roughly the size of Maryland.

In most situations, a flight plan describes specific flight lines that a pilot would follow, say from Airport A to Airport B, including where major turns would occur. Should that plane go missing, the CAP search initially would focus on the flight line plus a five-to 10-mile wide zone that parallels the planned flight track.

In this case, however, there was considerable uncertainty about the specific flight line. The only thing specified by the hunters was a general area, and some of this terrain might have been covered in a zigzag pattern. Given the large search area and the varied terrain, all possible search patterns were considered.

The CAP planning team laid out on large tables a series of aeronautical maps that cover the 13,000 square miles of terrain. The veteran CAP pilots then attempted to mentally place themselves in the seat of the missing pilot: Considering the various peaks and valleys, what would be the most likely flight line the pilot would have taken for this elk spotting trip? Soon, the team reached a rough consensus on the most probable flight line and it was marked on the map. It might as well have been in pencil, though, as everyone knew that this was an educated guess and simply marking the guessed flight line on a map doesn't make it real.

Within a few hours of being activated, the CAP had all of its 210 pilots and 100 aircraft on standby status, as was normal protocol. Al Lopez, however, could only utilize a fraction of these assets for safety reasons. The mountainous search terrain dictated extreme caution, as the thin air affects engine performance and there was the possibility for severe downdrafts developing. A limited number of planes had the lift characteristics suitable for handling these conditions. Also, the terrain made it more difficult to ensure adequate separation distances between the search planes, restricting the number of planes in the air. Even veteran CAP pilots underwent specialized training to fly mountainous terrain.

In addition to the CAP assets, the military and state police were put on standby, with potential requests for air support and personnel. Volunteer search teams from northern New Mexico were notified.

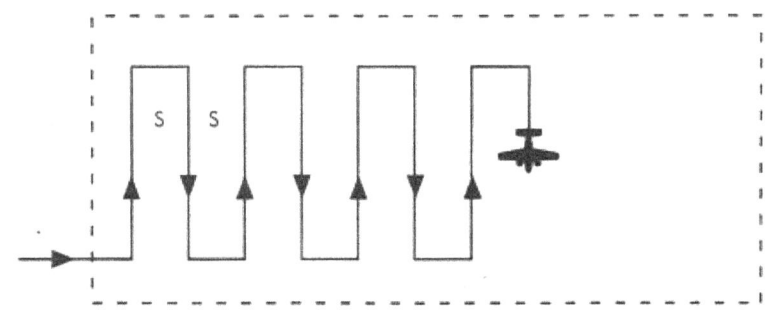

Parallel Track Search: This procedure is normally employed when the search area is large and fairly level, and the approximate location of the target is known.

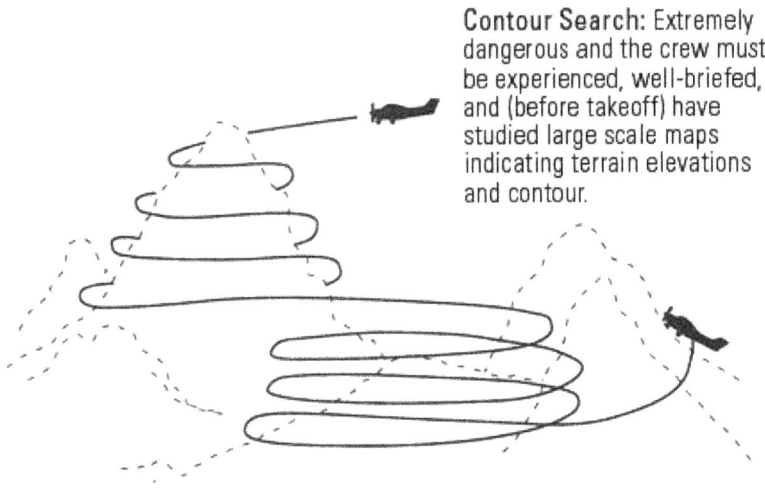

Creeping Line Search: This procedure is normally employed when the search area is narrow, long, and fairly level.

Contour Search: Extremely dangerous and the crew must be experienced, well-briefed, and (before takeoff) have studied large scale maps indicating terrain elevations and contour.

Had the hunters' plane gone missing a decade later, the initial search would probably be electronic, looking for a signal from an Emergency Locator Transmitter. Those searches are fast and can be performed in poor visibility and at night. However, in 1968 use of the Emergency Locator Transmitters was mainly limited to military aircraft operating in the Vietnam War. So, the search for the Fishel plane was conducted using only visual methods. Visual searches are slow and difficult.

Each search plane carried a pilot and at least one observer. The pilot's focus was on keeping the plane safely above the ground and away from other planes. The observer was trained to rely heavily on unaided eyesight to look for any out-of-the-ordinary signs. Occasionally, binoculars would help the observer, but their use was mostly to confirm initial sightings as the binoculars greatly restricted the field of vision (try looking through the side windows of a moving car using binoculars).

As challenging as visual searches were, they were successful due to one important factor: Geometric shapes are highly visible from the air because they don't occur in nature. A search craft can spot a vehicle or a person in open area from as much as two miles away. However, if the crash occurred in a wooded area, the visibility is a half mile or less. Often the observer first sees the debris field surrounding a crash site rather than the body of the plane. The most difficult to spot are the "corkscrew or spin" crashes, in which the plane descends in a near vertical angle and neatly disappears beneath the tree canopy with little debris field.

Rigid rules apply to conventional flying for all pilots. The pilots run through extensive pre-flight checks of equipment and crew to minimize the risk of error. It is challenging simply to fly a plane from point A to point B. The level of complexity, though, grows greatly when searches are conducted. Not only is the search crew faced with long hours, they usually are flying in terrain or in weather that may have caused a plane to go missing. With the potential for several things to go wrong during a search, CAP training is designed to break "The Error Chain.":

"A series of event links that, when considered together, cause a mishap. Should any one of the links be "broken", then the mishap probably will not occur."

It is up to each crewmember to recognize an error and break the error chain. Each sortie flown by search aircraft is planned and potential trouble areas are anticipated pre-flight. Even after going airborne, the crewmembers perform the *Stupid Check* to verify navigation. Do our headings, etc. look sensible? Any member of the crew may halt a flight with "Hey wait a minute. This is stupid."

It has been estimated that for each hour a pilot is in the air, 10 additional hours are required to put him there. This ratio considers the many behind-the-scenes people to support the flight. They include observer communicators, mechanics, drivers, ground search teams, and others.

Five planes took to the air the afternoon of the first search day, October 2nd. The planners assigned one plane to duplicate the entire projected route from Albuquerque over the Pecos Wilderness and Sangre de Cristo range to the Colorado state line and back. Four other CAP planes flew widely-separated peripheral areas with little tree cover, such as meadows, that could be rapidly viewed. If John Fishel had partial control of his aircraft, yet knew a crash was imminent, he likely would have directed his plane toward the open areas.

The search efforts "were without success," a CAP spokesman said. A fire seen in the Wheeler Peak area, possibly caused by a plane crash, turned out to be a prospector's campfire, the State Police reported.

Meanwhile, the CAP planners, sequestered in the adobe building at the Santa Fe Airport, worked into the night in anticipation of a full scale search at daybreak.

Day 2 brought 13 aircraft into the search, including an Army helicopter and another flown by the New Mexico State Police. The search, however, was called off early in the afternoon first by high turbulence from a fast-moving weather system, and later by rain showers. In a particularly devastating blow to the search, the storm dropped fresh snow in the high country, greatly hampering the search for the mostly white airplane.

Despite the setbacks imposed by the storm, the search received its first leads on Day 2. After seeing some of the news coverage, Pecos Valley resident Stanley Gonzales told State Police he saw an airplane in trouble over Dalton Canyon at about eight fifteen Wednesday morning. The Dalton Canyon area lies some seven miles northwest of the town of Pecos and due east of Santa Fe. Gonzales accurately described the plane as blue and white with tricycle gear, having one wheel under the nose and two wheels under the wings.

Curiously, the time of Stanley Gonzales' sighting was based not on a clock but on when the mail was delivered. The State Police officer listened patiently as Gonzales explained that he had just returned from setting up a hunting camp in an adjacent canyon when the plane flew over his house in Dalton Canyon. He then fed the dogs, went into the house to rest for a few minutes, and got up when the mail arrived predictably, as always, at five after nine. After considering how long it normally takes him to feed the dogs plus the length of his typical rest, he concluded that plane flew by fifty minutes before the mail, putting it around eight fifteen a.m. The searchers now had a location and a time marker to target.

The second potential lead came from a cook who worked at a country club on the eastern edge of Albuquerque. He saw a fireball and trail of smoke in the vicinity of the Sandia Mountains. Even though this sighting was some 75 miles away from the Dalton Canyon sighting, the searchers pursued both leads. The cook told the *Albuquerque Tribune* that he was driving to work when "I glanced up...it was about seven a.m. or five minutes before... and I saw this ball of fire." He said the object was headed north toward the south end of the Sandias and was below the skyline of the 10,000-foot mountain peaks. The next day, two Civil Air Patrol planes swept over the Sandia Mountains looking in the area described by the cook. No evidence of the plane or fire was found.

Early morning light on Day 3 saw two CAP planes headed for Dalton Canyon and nearby Thompson Peak, but they were forced back by heavy clouds that blanketed the wilderness. Meanwhile, Dr. Norris Bradbury, director of Los Alamos Scientific Laboratory, granted leave to employees who wanted to participate in the search.

Frustrated at having to idle their search planes because of foul weather, the CAP officials turned to organizing a ground search of the area. Time was critical, and the search couldn't stop while waiting for the skies to clear—let's get some boots on the ground and look. A large meeting to plan the ground search took place in Albuquerque, Friday afternoon. Twenty-six men plus the wives Sara Brittain and Beverly Horton participated. Among the groups represented were the Civil Air Patrol, EG&G, Gulton Industries,

Sandia Corporation, Woodman of the World, New Mexico Mountain Club, Seven Bar Flying Service, and Eastern Hills Baptist Church, where Ken and Sara Brittain were members.

Although the CAP had jurisdiction over the air search, they could not legally direct a ground search consisting of civilian volunteers. However, they agreed to suggest search methods and locations. Each group of volunteers would provide its own transportation, food, and shelter. It was agreed that volunteers would rendezvous six a.m. at the Santa Fe Airport before heading out.

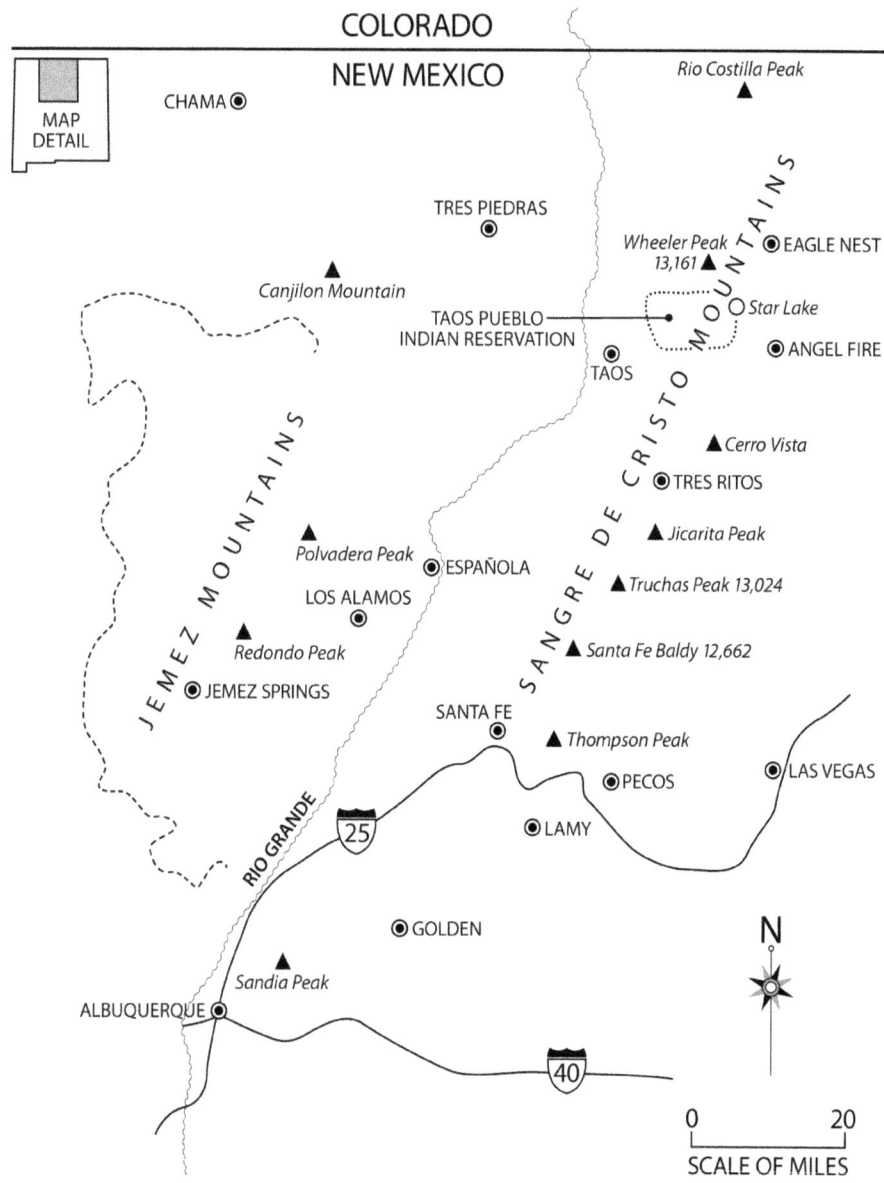

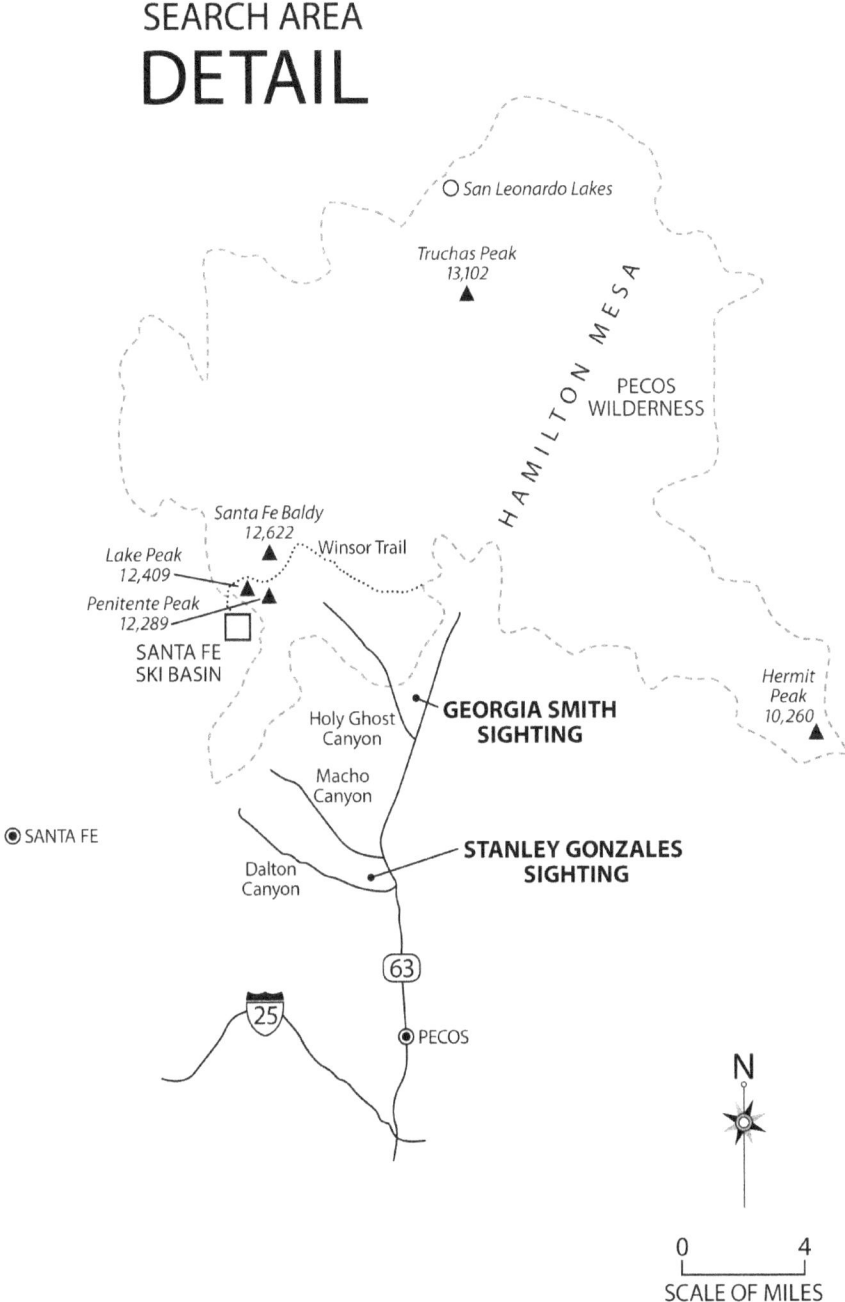

8

The massive ground search began on Saturday morning, which, coincidentally, was opening day of the elk hunting season. Expectations and anxieties were high for the ground search as friends and family of the four men finally could directly help find their loved ones. An estimated 200 people were participating in the organized ground search plus unknown numbers of others who headed into the mountains on their own. Gulton Industries was in charge of the ground search.

The ground searchers blanketed the southeastern portion of the Pecos Mountains. Particular attention was focused near Dalton Canyon, where the Stanley Gonzales sighting occurred, and a nearby drainage toward 10,540-foot high Thompson Peak. Among those searching were Sid Brittain and Bud Hedges, uncles of Ken Brittain, who brought their own horses from Phoenix, Arizona. Many of the volunteers were members of the Caravan Club, a hiking group of Santa Fe and Los Alamos residents.

Clearing skies Saturday morning meant that a full-scale aerial search could be conducted for the first time. Some 50 planes were mobilized from Albuquerque, Santa Fe, Los Alamos, and even from Alamogordo, located some 190 miles south of Santa Fe. The aircraft included Civil Air Patrol and private planes and three large Navy Grumman Albatross amphibious planes from Luke Air Force Base, Arizona., used by the U.S. Navy for low-flying search and rescue missions.

The search covered the area from Alameda Airport to the Sandia Mountains and up into the Pecos Wilderness. Some planes flew north to the Colorado Border. Coverage was intensive in the National Forest areas, particularly north to Tres Ritos. The Navy amphibians ran from Santa Fe Baldy to Albuquerque and along the Rio Grande. Other planes flew west to

Mt. Taylor and the Jemez Mountains above Los Alamos. Several leads were checked out and two of them turned out to be old plane wrecks, while another proved to be a pile of garbage and tin cans.

Maximum search efforts again continued on Sunday, the fifth day. Twenty-five planes and an U.S. Army helicopter from Colorado Springs participated.

A potential breakthrough: an object described as "possible plane wreckage" was sighted at six-fifteen p.m. by a ground party 2½ miles away, but the party was unable to reach the object because of approaching darkness.

Lieutenant Al Lopez of the CAP was faced with a difficult decision. Normally search efforts would resume the next morning to avoid concerns with crew fatigue and poor visibility. However, two considerations caused the CAP to consider an alternative. First, with the clearing skies, night temperatures could easily dip below freezing in the Pecos area, greatly stressing any possible survivors. Second, a full moon rise at six twenty-five p.m. would soon illuminate the landscape.

After discussions with the crews, the CAP reached a nearly unprecedented decision that planes would search the area of the possible wreckage by moonlight. Unfortunately, dense forests in the area prevented any sightings of the object from the air that night.

It took a team of expert climbers most of Monday to reach the possible wreckage site, guided by planes circling overhead. The climbers reported that the object was an aspen log gleaming in the sunlight.

Search parties and pilots were plagued by rugged terrain and deceiving shadows in the densely forested area. As CAP information officer Earl Livingston put it, "You're never sure of what you're seeing in the early morning and late afternoon shadows." Complicating these factors were patches of snow left by the earlier storm. Although the CAP efforts continued, Al Lopez hoped that one of the hundreds of elk hunters in the wilderness will stumble onto the wreckage.

Day after day, the CAP and the ground searchers worked together and pursued every credible lead they had. Every promising lead proved to be frustratingly false and the air search would soon end. As a final measure, on Day 8 a specially modified F-4 fighter jet was deployed from Bergstrom Air

Force Base near Austin, Texas to fly a high-altitude photo reconnaissance of the search area. The photos were processed the next day.

Meanwhile, The Air Force's Central Air Rescue recommended Tuesday, October 8, that the CAP search be moved to the west side of the Sangre de Cristo mountains and along the Rio Grande valley to the Colorado state line. The CAP said the absence of any concrete leads indicates that the Piper Cherokee may be down in a heavily-timbered area where it would be difficult to see from the air.

While Beverly Horton focused on the ground effort, Sara Brittain turned her energies to writing a series of letters to government officials. Concerned that the CAP would soon stop its search, Sara wrote to Joseph M. Montoya, senior U.S. Senator from New Mexico. She urged him to encourage the CAP to continue their searches. The Senator quickly responded:

> "...I have contacted officials of the Civil Air Patrol and have urged that they continue every possible effort in the search. I have commended the untiring efforts of the Civil Air Patrol in conducting this search. ...The Civil Air Patrol has assured me that they will continue to check on any possible lead or report which they receive on the missing aircraft...."

Another letter written by a family friend went to Governor David F. Cargo and requested more involvement by the New Mexico National Guard. Cargo responded two days later and said that he was unable to utilize the Guard on full-time basis because of the "federal aspects involved in paying for the personnel." He assured that he was following the search very closely and "will do everything in my power to see that these men are found."

9

The one constant in the lives of the wives was their husbands. No matter which city or part of the country they lived in, their husbands were their soul mates and confidants. Sure, they led busy lives with school, volunteering, or parenting. However, it was their time together as a couple that nourished their roots. Whether it was attending church together or family camping trips, the couples relied on each other and grew from their common experiences.

For many, 1968 was a tumultuous time marked by war, assassinations, civil unrest and deep political division. At the same time, it was a much simpler time for families than today. The three couples maintained normalcy through their reliance on each other. No doubt that it was a man's world in 1968, yet the men got strength from the daily contact with their spouses.

All of this fabric was ripped when the wives received early news of the plane's disappearance. At first, the event seemed quite unreal and the wives remained optimistically naïve that their men would simply walk through the door that night. At the vibrant ages of 27 to 29, they did not accept the possibility that their spouses would not return. The first few days after the disappearance, the wives stayed home, near the phone, at the request of the CAP. With frequent visits by friends or the continual need to deal with the kids, they tried to keep busy and anxiously awaited any news from the CAP. Daily, the wives got a call from someone in the CAP about the status of the search. The numerous false leads, however, caused mental numbness.

Beverly's daily routine since October 2 had been drastically altered. She quit her classes at the University of New Mexico and spent much of those first days on the phone. The radio mostly stayed off because the CAP cautioned that they could hear erroneous reports and get their hopes up for nothing. The love and companionship of relatives and friends gave her the strength to face the long nights and days of uncertainty. Jon Dale's parents, Mr. and Mrs. John Horton, drove to Albuquerque from Dallas the day after the plane was reported missing, and moved in with Beverly.

Like Beverly, Sara was immediately visited by Brittain family members. On Day 3, two of Ken's uncles from Phoenix arrived with their horses to help the ground search. Shortly thereafter, Ken's parents, siblings and cousins arrived.

On the third day of the search, Sara and Beverly went to meet Barbara Jones and the newly-formed alliance of the wives helped some. Still, the nights brought little sleep. The initial optimism was being replaced with discouragement.

Sara's daily entries in her dairy showed the peaks and valleys of emotion. On day two: "Seems unreal. We hope and pray they're all right and will be found." On day five: "Felt so lonely and discouraged tonite. They're working one lead. We want them to be found, but are afraid too." On Day 6: "Stayed home. Lead was a dud! I feel numb—hard to express grief openly. It just doesn't seem right to not find them, know story. Search continues tomorrow." On day seven: "Nothing again. It'll be a week tomorrow. We'll check on clue from man who saw airplane last Wed. We grasp at anything and will hope to the last!"

Of all the wives, Barbara Jones could have felt the most isolated and alone. Since the day of the flight, she progressively became more depressed and convinced that Ron was dead. The difficulty of dealing with the uncertainty was amplified by her isolation from family in Florida. She was visited by Ron's parents, but the travel was hard for them. Sara and Beverly were supported extensively with family in states adjacent to New Mexico, but Barbara had no real ties to the area. After her belief that Ron was gone manifested itself, she began to think of returning to Florida. Deciding that she didn't have it in her to physically join the searches, she hunkered down at home with her kids and awaited some news.

It is an inherently human characteristic that we don't know ahead of time how we'll react to our first life-changing crisis, such as when a loved one unexpectedly dies. Some people suppress or control the emotion of the moment and shift those emotions to take on the challenges that lie ahead, like organizing a funeral. Others are so overcome with grief that they cannot immediately respond. Some people with missing family members say they learn to think about the missing person in the abstract, so they can better deal with it.

For Sara Brittain and Beverly Horton, a slow transformation began to take place after the first few days of sitting around in their houses awaiting news.

A steely determination slowly set in to do whatever they could to find their husbands. Without consciously intending to, they would become a bonded force that would prod, push, or guilt the media, government officials, and the public into not giving up or forgetting. Their initial feeling of helplessness would be replaced with a drive to find the answers. Choosing to be optimistic and hopeful, they moved into action. No longer would they, or could they, stay in the background. The more they kept busy, the less likely they would be to succumb to the paralysis brought on by blue periods.

An opportunity for involvement presented itself to the wives the first week. An Albuquerque group of individuals and organizations offered to put up a $500 reward to the person who gave a definite clue leading to the whereabouts of the blue and white plane. Contributions were received from Gulton Industries, EG&G, Eastern Hills Baptist Church, and various friends of the families. "We are grasping at straws," Beverly told the *Albuquerque Journal*. "We have to have something … (W)e hope the reward gets people to recall seeing the plane, or to get local people out to help."

Sara, Beverly, and the Horton parents traveled extensively in the Tres Ritos, Cuba, Jemez Springs, and Pecos areas, tacking reward posters where hunters were most likely to see them. They visited with Boy Scouts troops and asked them to take the posters home and share with their parents. Occasionally, they rode in planes or helicopters to help look.

The Horton parents took an active part in the ground search while other relatives stayed near the house to help answer the many phone calls. Mr. Horton committed to remain in Albuquerque until a definite clue develops. "We'll stay until we find something… (O)ther people are working my farm for me now," he said. Mr. Horton often drove his camper to a search area and provided drinks and snacks for searchers.

10

The staff of Gulton Industries re-convened in their Albuquerque offices on Monday, October 7, after the massive ground search over the weekend. To a person, it was unacceptable that their colleagues were still not found. The company went into battle mode. They applied their considerable engineering and research skills to gathering and analyzing information.

They converted their largest conference room into a war room. Maps were hung from every wall and peg tacks located details about the plane and the search. It was not unusual for the staff to work a full day on the search, with their time fully paid for by Gulton, only to return to the office later to keep up with corporate commitments. Ed Whaley dedicated himself to directing the ground search. Somehow the company would meet all of their delivery deadlines to corporate customers, he trusted.

One of the first technical issues to evaluate was the expected performance of the airplane. It is likely that the CAP had already evaluated this, but the engineers felt the need to understand the situation better. Three phone calls on Tuesday morning, October 8 provided sobering information, confirming what many had worried about.

The first call was to the manufacturer of the rental plane. They were told that with a new engine under idealized conditions of air temperature, the Piper Cherokee could operate to a ceiling of 16,400 feet. That altitude, of course, assumes the pilot can function in such a low oxygen environment because there is no supplemental oxygen on the plane. However, because of a weather front that entered the area that day, the plane would have been subjected to mild turbulence, lowering the operating ceiling from idealized. Affecting the performance further was the fact that actual air temperatures on the flight day were some 20 degrees F warmer than the idealized standard temperature. These factors would reduce the ceiling by about 3000 feet, bringing the service ceiling to 13,400 feet maximum. Under these adjusted conditions, the manufacturing rep estimated that cruise speed would barely reach 140 mph, and the rate of climb would be slower than normal. Contrary to what most of us would think, warmer temperatures actually worsened the performance of the Cherokee. After adding the weight of four 200 pound men, it is clear that the plane was being pushed near its limit while flying around mountains that top 13,000 feet.

Based on the flight plan and the predicted flying speed and fuel reserves, there wasn't any extra time for the hunters to look around, according to the sales rep. In fact, he believed there barely was time (two hours) to fly over each announced peg point and return home.

One hour after talking with the Piper experts, the staff at Gulton received a phone call from Bob Sylvester of Sandia Corporation. Sylvester was a pilot who was familiar with the Piper Cherokee 180 and had some observations to offer the search team. He started by offering the opinion that a practical ceiling for the plane was 12,000 feet, particularly since the plane did not carry oxygen. Sylvester went on to review of the weather conditions on the flight day Wednesday, October 2nd. He said there was a front moving in from the northwest, with winds out of the northwest at about 25 knots at 10,000 feet in the area of question. The effect of these winds, depending on the slope of the ridge, would be to cause downward vertical currents of about 2,000 feet per minute on the leeward, or downwind, slope. He estimated that the plane with its load had a vertical rate of climb of about 250 ft/min. If the plane were caught in a downdraft, it would fall at the rate of 1,750 ft/min vertically.

There is great concern of being caught in a downdraft when flying anywhere near a ridge line. Even if a plane is initially flying above the ridge line, it still could be caught in downdraft winds as the plane nears the ridge crest and then sucked downward into the leeward slope. Sylvester concluded that search should be made only of the eastern and possibly southern slopes of the ridges. These were the downdraft slopes based upon the estimated direction of wind—that was out of the northwest.

The third call that morning was to Major Findley of the Civil Air Patrol. The CAP drew similar conclusions about the likely performance of the airplane and they thought it was unlikely that the plane reached the furthest boundaries of the search area. In addition to limits on engine performance at this altitude, the Major mentioned that it did not appear the plane's fuel supply was topped off. He estimated the plane departed Albuquerque with 30 gallons of fuel instead of 50 gallons in a full tank. As a result, the plane may have departed Albuquerque with only two hours of fuel on board, instead of the 3½ hours mentioned in the verbal flight plan.

The massive ground and air searches that took place the previous weekend focused on the Dalton Canyon area. Many believed that Dalton Canyon and adjacent canyons still seemed to have the greatest chance for holding the plane, based on the wind characteristics and the Stanley Gonzales sighting. The information gained from these recent phone calls only strengthened those beliefs. Unfortunately, no plane was found in Dalton Canyon or in other leeward canyons, despite intensive searching with fixed wing aircraft, helicopters, and hundreds of ground searchers.

The Gulton team decided to direct the next large ground search to the Tres Ritos area—the location that Jon Dale Horton and Ken Brittain had selected for their upcoming elk hunt. Details of the Tres Ritos search would be finalized at a mid-week gathering at Gulton Industries.

As Ed Whaley opened the meeting, he must have felt a flush of gratitude and pride in his company: a total of 59 staffers had volunteered for the ground search. Every facet of Gulton's Albuquerque offices was represented. There were 34 volunteers from Engineering. The Machine Shop had six. Production had nine. There were others from Overhead, and Plastics, and Sales.

There were several pickups offered for the search, plus a Corvair, a Jeep, a couple of VWs, a Plymouth station wagon, a Falcon, and a Ford Fairlane. The list was rounded out with a touch of class: a '57 Ford.

Ed Whaley spent three days organizing the ground search. Somehow the whole planning exercise seemed somewhat mechanical and logical. He considered all the data before deciding where to search, then recruited and notified possible searchers for all the various groups, and finally he assigned searchers to a designated area. Because of this mechanical process, he could remain dispassionate and removed from the human elements.

On October 11, the day before the Tres Ritos ground search was to start, Whaley had a much more personal task at hand. He drove to Sara and Beverly's homes and delivered their husbands' payroll checks. In an extraordinary gesture, Gulton Industries would continue to pay Ken Brittain and Jon Dale Horton salaries for three months after their disappearance.

Ed Whaley then turned the discussion to a much more delicate matter: an overview of the Gulton life insurance policies each held. However, the discussion was brief, as Sara didn't think she could discuss this "until she knew for sure what happened to Kenny." Whaley would return later with more specifics.

The insurance discussion with Ed Whaley shook Sara. After he departed, the day only got worse for Sara when she learned that the CAP had suspended their search.

On October 11th, after ten days of searching, the CAP announced that it had exhausted all leads and it was suspending its organized search. Although the families knew the decision was coming, their spirits were shaken by the news. The CAP search was one large and organized activity that gave them comfort simply because of its presence. Take the CAP away, and any airborne searches had to be undertaken by families and friends as funds allowed. CAP rescuers had spent 538 hours of flying time seeking the plane. The 10-day search involved nearly 500 people and more than 200 aircraft in this humanitarian effort.

Shortly thereafter, the wives of the missing men wrote to thank the Commander of Bergstrom Air Force Base for the F-4 photo reconnaissance flight.

"Although the photos taken by the Reconnaissance airplane did not turn up any leads, your effort has been greatly appreciated...." They went on to gently prod the commander for any possible added assistance. "We know that officially you cannot take further photos of the area. It is our hope and prayer that perhaps some training missions might happen to be in the area."

PART III

The Families Take Over

11

The ground searchers met in Albuquerque at five a.m. on Saturday morning, October 12th, and caravanned to Tres Ritos where they were joined with searchers from EG&G, Eastern Hills Baptist Church, and numerous other organizations. The searchers were led by Frank Reeder of Gulton Industries and John Sheehan of EG&G. The Forest Service put their people on alert for anything that looked out of place. The Sherriff's posse from Las Vegas, New Mexico joined in the search as well. Most people camped near the village of Tres Ritos, but Ed Whaley and four other Gulton Industries engineers backpacked in to more distant spots in the wilderness. For safety purposes, every volunteer followed a careful check in-check out procedure and were advised to carry adequate food, water, and emergency gear. If the plane was found, the volunteers were to immediately notify authorities. The search party was joined in the morning by Sara and her extended family. Jon Dale Horton's father and mother provided drinks and snacks for the searchers.

The weekend search failed to discover any signs of the plane or passengers.

12

After his brief attempt at explaining life insurance benefits to Sara, Ed Whaley realized that the wives needed some professional guidance to navigate the complex world of insurance and finance. He turned to Bill Carpenter who was director of Human Resources at Gulton Industries. Carpenter and Whaley knew that it would be difficult to get an insurance company to pay out until the plane was found.

Their initial fears were confirmed when they phoned the insurance company. Part of the conversation went like this:

"How can you be certain the men didn't just fly to Cuba?" the insurance man asked.

"Cuba? The Country?"

"Yes."

"What does Cuba have to do with this?

"The hijackings."

"What hijackings?"

"The recent series of plane hijackings to Cuba."

"Oh, yes."

"Well, they have gotten us sensitive to the possibility.

"Sensitive."

"Before we could pay out, we'd investigate this."

"Of course."

"Also, how do we know they just didn't fly somewhere to get away from their wives or girlfriends? Maybe Mexico."

"That's not possible. All of the husbands were good family men and they never would've gone along with such an idea."

"What about the pilot?"

"We don't know much about him. But the others we *do* know and they would never..."

"Still, we'll need to look into this."

Bill Carpenter researched all things financial for Sara and Beverly. Then, in a lengthy letter dated October 18th, Carpenter laid out the details of how to cope with the legal paperwork. He first dealt with the insurance that would come from Gulton's life insurance policy. It would be a one-time, non-taxable payment. The insurance company would verify the specific payment amount when a claim is filed.

Here is an excerpt of what he wrote to Sara:

> "If the plane is not found soon, Gulton should submit a death claim form with newspaper clippings as circumstantial evidence. (I will do this this week.) Because of the large amount, _____ (the insurance company) will begin an investigation as soon as the claim is submitted. If a death certificate cannot be obtained, (will not be issued until the body is found or until seven years have passed) (the insurance company) will arrange through their own bonding company to bond the policy. This bond is a reimbursement bond in case Ken is found alive. The cost of the bond will be borne by the beneficiary, and it varies from $6.00 to $7.00 per thousand. (The insurance company) will pay the insurance after the investigation is complete and the policy has been bonded. ...I will submit the claim and contact you if bonding is necessary."

Carpenter then summarized the amount in Ken's savings account, held in the company's credit union. As with the life insurance, all savings in the credit union were insured for double the amount in savings. Based on conversations he had with the local credit union office, if the plane is not found, they would probably pay after one year's time. "We will follow this further and it will be taken care of at Gulton."

The next topic was benefits to be owed by the Social Security Administration. Payment would be based on whether the wife has children, the age of the children, and whether the wife works.

"Without proof of death, circumstantial evidence (newspaper clippings) is required. It is very important that the claim be submitted promptly. Even if the circumstantial evidence was not accepted, and no benefits were paid until after seven years, if the claim were filed now, all benefits for those seven years would be payable. If you wait seven years or until the plane is found to submit a claim, you receive none of the retroactive benefits. Benefits are retroactive to the date the claim is processed."

Bill Carpenter closed his letter out with information about taxes. He mentioned that an accountant in Gulton's Data Systems volunteered his services in preparing their income tax forms for the coming year. Finally, in a hand-written postscript, he urged the wives to talk to an attorney later on about drawing up a will for themselves.

The upshot of all this research was that the wives faced an uphill battle to get the insurance company and the Social Security Administration to pay off without a death certificate or waiting for seven years of time. Thinking in these terms was alien for the wives. It was in stark contrast to their belief that their husbands could still be alive out there—somewhere. The world of death certificates and double indemnity insurance policies were a galaxy away from their lives a few days earlier. Nonetheless, with the care of the experts at Gulton, they would be forced to use newspaper clippings to argue what they didn't want to argue or believe—their husbands were not coming back.

Beverly and Sara decided to go ahead with submitting claims for Social Security survivor's benefits. True to his previous efforts to help the wives, Bill Carpenter of Gulton accompanied the wives to the Social Security Office on October 24th. Ostensibly he was there in case they encountered unexpected questions or difficulties. Mostly, though, his presence was a comfort to the wives. The visit to the Social Security Office took place three weeks after the plane's disappearance.

13

On October 22nd, Ed Whaley received a peculiar phone call. The caller identified herself as Mrs. Harold E. Mott, and a close personal friend of the famed psychic Jeane Dixon. The Motts had homes in Washington, DC and Window Rock, Arizona; the dual residency was necessary because she worked in some capacity with the U.S. Government and the Navajo Nation. Mrs. Mott was in New Mexico doing some work with the Navajo Indians and had heard some of the news coverage about the missing plane. She said Jeane Dixon probably would be willing to help out. Would Ed Whaley need her help?

Psychics, clairvoyants, seers: Whatever label is used, humans have been interested in the sixth sense since ancient times. The 1960s and '70s brought resurgence in interest in ESP that probably is unequaled in modern times. Daily, stories could be found in the news media about alleged breakthroughs in telepathy, psychic healing, or prophecy. Three of the most famous psychics of the time were Jeane Dixon, Peter Hurkos, and Dr. Gilbert Holloway. Each of these eventually would play a role in the search for the plane.

Jeane Dixon was the psychic most well known to Americans. She became part of our life as a regular contributor to Parade Magazine, the popular weekly insert found in the Sunday newspapers. Her apparent prophecy that President John F. Kennedy would be assassinated propelled her to fame. In the May 13, 1956, issue of Parade Magazine she wrote that the 1960 presidential election would be "dominated by labor and won by a Democrat" who would then go on to "(B)e assassinated or die in office though not necessarily in his first term." Her annual forecasts of the future in *Parade Magazine* were highly anticipated and entertaining.

That many of her other predictions were not as accurate failed to diminish her popularity. Critic James Randi in his book *Flim Flam* mentions a few of her less than stellar predictions for 1978: President Jimmy Carter would resign (false); big-haired actress Farrah Fawcett would get a crew cut (false); Pope Paul VI "would astonish the world with his vigor and determination" (he didn't live out the year). Mathematician John Allen Paulos of Temple University is credited with coining the phrase "The Jeane Dixon Effect," which refers to the tendency to focus on a few correct predictions while ignoring a larger number of incorrect predictions.

Dixon's public stature grew through the 1965 biographical volume, *A Gift of Prophecy: the Phenomenal Jeane Dixon*, written by syndicated columnist Ruth Montgomery. The book sold more than three million copies. She attributed her prophetic ability to God. Her influence was profound in many circles.

Ed Whaley did not immediately respond to the Jeane Dixon possibility. From nearly the first day of the search, he and others had received many unsolicited 'readings' of numerologists and clairvoyants. While he personally was skeptical of the readings, he treated each one with proper due diligence, as a trained engineer would. He later said,

"You know there were a lot of clairvoyants that called in, maybe a half dozen of them. I took anything that a clairvoyant said just as if a guy said 'I was standing there and a plane flew over.' I took it just as valid as any information, until two of them put it in the same place. (I was skeptical) because they had the plane up in San Antonio Mountain (at the Colorado/New Mexico border), they had them in Tijeras Canyon (southeast of Albuquerque), and even out in the Mt. Taylor area (60 miles west of Albuquerque). They had them scattered all over the bloody State. So I just took them and tried to write down what they said.

"I would accommodate them as I would any other sighting or any other information. In other words, we wouldn't just throw it in the file and forget about it. We would say "Let's follow up. Do we have any other data that would support this?" If someone said they were over at San Antonio Mountain, we'd look at the facts: We knew what time they left, how far away that was. How many hours would it take to get there, and we know what time they were supposed to be back, etc."

Even Whaley acknowledged, however, that this call was different because of Jeane Dixon's fame. He went ahead and passed Jeane Dixon's contact information on to Sara and Beverly.

14

As pastor of the Eastern Hills Baptist Church, Rev. C.E. Allbritton took great pride in the cohesiveness of his congregation. They looked out for one another. He considered himself especially lucky to have among his congregation Sara and Ken Brittain. They gave selflessly of their time. On any given day, chances were high that Sara would be playing piano for the service.

When word spread that Ken Brittain was missing, the church members immediately wanted to help. They became main fixtures of the ground searches and helped establish the rescue reward fund. Initially, Rev. Allbritton oversaw the church's search activities. In short time, however, he realized that Ed Whaley and the engineers at Gulton Industries were doing a capable job and he decided to defer to them. Henceforth, he was a regular feature at any planning meeting held at Gulton Industries.

After the disappointment of the Tres Ritos ground search the past weekend, Rev. Allbritton was excited to pass to Ed Whaley what seemed like credible leads. A church member was hunting in Pecos and saw a large number of birds (vultures, etc.) hovering at a spot. The birds could have been after the remains of an elk, but they also could be at the site of the downed aircraft. The pastor asked Whaley's staff to follow up on the lead.

Further checking by the Gulton staff determined that the birds were located on a rock slide above the Pecos River and across from Hamilton Mesa. This got their interest because Hamilton Mesa was one of the hunting locations that EG&G employee Jack Frost had recommended to John Fishel. A ground search party led by Morys Hines of Gulton hiked to the site that day and found a dead bull elk. "Rather relieved," Sara wrote in her dairy that evening.

A second lead passed on by Rev. Allbritton involved the sighting of a tail and door of a plane. This proved to be an older plane crash. The third lead in two days came to the Gulton Team as a sighting of "shiny metal", but a helicopter was unable to get to the site. A ground party was to go in during the coming weekend.

While false leads came to the searchers from day one, they still were encouraging to the families. The leads showed that the public hadn't forgotten about their loved ones. The wives announced that the reward fund now stood at $1000, in hopes of keeping the public interest. Nonetheless, doubts crept into their heads more and more often. Before resting that evening, Sara wrote "I'm afraid and fearful for Ken's life."

While en route to investigate the gathering of birds, Morys Hines and two others from EG&G stopped in to visit Stanley Gonzales, the rancher who reported seeing the plane fly over the trees on October 2nd. His wife said that Stanley was out searching for the plane with the local Sheriff's Posse and a couple of State Police Officers.

The group returned to the home later in the day and they were greeted instead by Georgia Smith, friend of Stanley Gonzales and caretaker of the Hidden Valley Campground in Holy Ghost Canyon. After some brief introductions, Georgia Smith offered that she had something to share with the group that might be helpful. Georgia told the searchers that early in the morning of October 2nd she saw a plane go low over the trees in a northeast direction at the top of the bluff directly behind her house. This bluff is also behind Stanley's house. (Holy Ghost Canyon is a major drainage north of Dalton Canyon.) She said that there was lots of sound—possibly full throttle.

This chance encounter bolstered the belief that intensive searching of Dalton Canyon and Holy Ghost Canyon was appropriate. Two possible sightings of the missing plane in the same vicinity, a few minutes apart.

Once the Civil Air Patrol disbanded the search, the responsibility fell to the families and friends. Searches from the air were expensive and dangerous, but remained the most efficient way to scan thousands of square miles of landscape. Several private pilots generously volunteered their skills and even their own aircraft to continue the search. Hardly a day elapsed when there wasn't a team of volunteers aloft looking for the missing plane. As long as the pilot had a spotter to accompany him, he flew regularly. On more than one occasion, the pilots were forced to quickly retreat to Albuquerque on the onset of severe weather.

From Gulton Industries, mainstays were Dave Wirenius and John Wheatley. Wirenius devoted a month of his and Gulton's time to the effort. For the first week he flew as a spotter with a helicopter crew. Then he switched hats and piloted his own plane wherever the Search Team directed. Occasionally his spotter was George Friberg, a manager at Gulton's semiconductor manufacturing plant and close friend and fellow church member of Sara and Ken Brittain. On one particularly turbulent flight, Friberg commented to Wirenius that this may be one time that Gulton would have to rescue the rescuers. Then he turned away pale, and proceeded to throw up on his letter jacket received from his early days as quarterback at the University of New Mexico.

Frank White and Glenn Todd from EG&G would hurry to the airport after work and fly until dusk, as well on the weekends. Joe Hoffman, the technician who worked side-by-side with John Fishel and Ron Jones, was a regular spotter. Ed Young from Sandia Corporation also helped. In addition to providing flying time, he gave valuable technical insights to the Search Team because he also flew a Piper Cherokee aircraft.

15

Stunt pilot Doug Rhinehart was keenly interested in all matters related to airplanes. He closely followed in newspapers and on television the search for the missing Albuquerque aircraft. Although he lived hundreds of miles away in the Four Corners section of New Mexico, the extensive news coverage kept him apprised of developments in the search. It was in an *Albuquerque Journal* story about Beverly Horton that he read the CAP had suspended the search. He decided he needed to do something to help. It was an unwritten ethos that fellow pilots help others when possible.

He contacted Beverly Horton and volunteered his time and plane. Now that all the CAP aircraft were gone, he thought that he could find the missing plane. He explained to Beverly that he wasn't rich, and he had a family: Could she cover his expenses such as food, hotel, and fuel? Fortunately, Beverly had recently been given a gift of $2,000 by the Albuquerque Boat Club (Jon Dale and Beverly were members) to use anyway she wanted. Beverly told him she could give him the $2,000, but she wanted him to fly until that ran out.

For Doug Rhinehart, flying was one of life's gifts. As a stunt plane pilot, he could wow people. His little canvas Rose Parrakeet bi-plane could defy gravity. Upside down! Loop! Spin! Fly to the sun, kill the engine, and glide toward the ground! The plane's diminutive stature belied its performance: The single-seat plane was not as tall as Rhinehart and the fuselage was barely 16 feet in length. With a short wing span and exceptional lift provided by the two wings, the fixed-wing plane had agile handling characteristics that could rival that of a helicopter and was often used for acrobatic performances.

In truth, it was because of the little plane's maneuverability that he was alive at all, according to a story he shared with Beverly Horton. One time he encountered some violent weather while flying in some mountainous terrain. The severe winds forced him to make an emergency landing in a remote canyon, but because of the plane's ability he was able to safely set down. By the time the storm passed, it was dark. Through the night he collected dew drops off the plane's wings for drinking. With morning's light, he saw that the plane was airworthy and he flew out.

Rhinehart seemed destined as a kid to become a pilot. In his teens during the 1930s, he worked as an airport boy and learned to fly at Roger Airport in Springfield, Missouri. It was there that he saw his first Rose Parrakeet; it was an event that planted a seed. Three decades later, in 1963, Rhinehart finally fulfilled a life-long dream and bought a 1938 Rose Parrakeet. This placed Rhinehart in unique company, as he owned one of only nine of the planes manufactured by designer Jack Rose. Rhinehart's love of the plane wasn't satisfied yet. In 1965, he purchased two more of the original Rose Parrakeets—also from the 1930s—and had obtained permission from Jack Rose to begin manufacturing five additional planes using Rose's design specifications.

The purchase of his first Rose Parrakeet allowed Rhinehart to begin flying as a stunt pilot. When Rhinehart wasn't flying in shows around the country, he lived near the Four Corners in Farmington, New Mexico. He located the Rhinehart-Rose Manufacturing Hangar in nearby Aztec, New Mexico.

Two days after his phone call to Beverly Horton, Doug Rhinehart and a Rose Parrakeet nicknamed "Ramblin Rose" arrived in Albuquerque on October 19th. He housed the bi-plane in a friend's hangar at the Alameda Airport. Ironically, the airport was the same facility from which the hunters took off. Beverly and Sara had dinner with him at his motel's coffee shop and discussed where he should fly. They decided to initially target Dalton Canyon and Holy Ghost Canyon.

Alone after dinner, Rhinehart took some time to review what he knew about the situation. He mentally worked through the obvious problems the hunters could have encountered. One of the problems he considered was whether the pilot of the Piper Cherokee could have been blinded by what the pilots call "sun in eyes." Laying out his Aeronautical Chart of New Mexico, he sketched out the hunters' probable flight path from the Alameda Airport toward the town of Pecos. Calculating where the plane should have been every 10 minutes after departure, he tracked where the sun would have been relative to the flight path. He concluded that the crash would have occurred in the first 40 minutes of flight if the sun was a factor, in which case the most likely spots would have been in mostly open range lands before the big mountains of the Sangre de Cristos, and wreckage should have been easily spotted. The "sun in eyes" concern was dismissed as unlikely, and he reaffirmed to himself that their focus on the mountains was appropriate.

To maximize the time he could spend flying, Rhinehart wanted to base operations out of an abandoned dirt landing strip located south of Santa Fe in the postage-stamp sized town of Lamy, rather than fly out of the Santa Fe municipal airport. This would place him about 10 miles closer to the target canyons. A problem, however, was there were no support facilities at the landing strip. To Rhinehart, this was a minor problem. No worries! Beverly and Sara would serve as the airstrip's crew, he convinced them with his winning smile.

The next morning Beverly picked up Sara in the blue van, drove to the motel, and delivered Doug Rhinehart to the Ramblin Rose. The wives then headed to Lamy, where the deserted airstrip was located next to the highway on a private ranch. They explained to the ranch owner what they wanted to do and asked permission to use the airstrip. It was granted, and by then Rhinehart was flying overhead waiting for their signal to land. No cell phones, no CB radios, no radios of any kind in the plane: All communications were done by hand and arm signals.

The blue van was conveniently equipped with a rack of army surplus gas cans mounted across the back bumper. Every day, Beverly and Sara loaded up Jon Dale's van with five-gallon jerrycans of fuel, and would make the trek from Albuquerque to the airstrip.

Because Rhinehart was not familiar with the search area, he prevailed on the wives to escort him on some of his flights: he wanted to be sure that he was searching the correct canyons. So, to enhance the escorts, the wives taped a big white arrow on top of the blue van, and then guided him into the different canyons. They also used the big arrow to give him wind directions on the airstrip. Periodically during the day, Beverly and Sara drove to the Santa Fe Airport for extra fuel.

After donning his leather flying helmet and extra insulation, Rhinehart took off in Rose Parrakeet and followed the van as they drove to the mouths of Dalton and Holy Ghost Canyons. He and his plane resembled "Snoopy" with the little red plane, down to the cap he wore. Once he fastened the ear flaps over his ear muffs he couldn't hear a thing, and the hand signals paid off even on the ground.

On late search days, Rhinehart buzzed the van and the wives quickly grabbed flashlights to illuminate the landing strip. Another time, Rhinehart was unable to land at the airstrip due to high winds and he wanted to set down instead at the Santa Fe municipal airport. Without any voice communication system with the plane, somehow Beverly understood what was needed. She drove directly to the Santa Fe airport, gained access to the tower, and told them that a stunt plane needed to land there. The tower granted clearance immediately and then, half-jokingly, asked her if he would do some stunts before landing. As if he could read minds, Doug Rhinehart directed the Rose Parrakeet through two rolls before touchdown.

Rhinehart believed he could find the plane in the first day, with his ability to fly low over just about any terrain. On Rhinehart's first day of searching, October 20th, he spotted plane wreckage and reported the find. The evening newspaper heard of the sighting and excitedly reported a possible breakthrough in the search. Unfortunately, the following morning the ground search team determined the wreckage was from an older plane crash. Rhinehart continued to look in Dalton and Holy Ghost Canyons, to no avail. With time, he expanded the search area, again to no avail.

Nine days after his arrival, Doug Rhinehart called Sara and Beverly to say that he didn't know how long he'd be able to continue. Sara's October 27th entry into her diary summarizes her reaction: "This is the worst blow I think I've felt since this started. Just can't stop trying!" The ladies, though, were deeply grateful for his help. Throughout his stay, Beverly was tickled by his sense of humor and felt he kept the mood upbeat. Beverly Horton was now five months away from delivering her and Jon Dale's first child. In the ultimate compliment to Doug Rhinehart's humanity, Beverly decided that her child would be named Douglas in honor of the stunt pilot.

16

Initial news coverage of the plane's disappearance was intense. Large front-page stories were printed in both major Albuquerque papers during the first week. Syndicated UPI or AP stories were picked up by other newspapers across the state and the region to broaden the coverage. Frequent radio and TV spots added to the mix. As happens with any news event, though, coverage had a finite life. After four or five days, stories in the papers began a march off the front page to the New Mexico or Local news sections. It wasn't as if the coverage disappeared: During October, there was some mention of the plane in *Albuquerque Journal* pages every day of the month but one.

As helpless as the families felt about many aspects of the plane search, one area they could affect was publicity. After the CAP suspended search operations, the families believed that their best hope for finding the plane was with the public. Keep the public looking and eventually the plane would be found, they believed. Consequently, the families and friends took every advantage of opportunities to generate publicity.

Thrust into the limelight with all the coverage, they quickly became wizened in the ways of public affairs. They did not hesitate to utilize the many media contacts that, sadly, they now had. For example, one opportunity unexpectedly came their way with the arrival of Doug Rhinehart, the stunt pilot. Prompted by a well placed phone call, a TV news crew arrived to interview Rhinehart and the wives. Newspaper photographers also magically appeared for the event.

Another opportunity was generated by Beverly Horton. She figured that reporters didn't hesitate to call her at any hour, so why couldn't she return the favor? Two days later, a four-column front page story about the search and how she was doing appeared in the *Albuquerque Journal*.

"Mrs. Jon D. Horton is a tall brunette who survives daily in her modest

Northeast Heights home on optimism and hope. 'I'm very optimistic, but can't say I'm prepared for anything...how do you prepare for things like this? ...I can't believe he is not alive,' Mrs. Horton said of her husband."

Although it was difficult for her to openly discuss her husband's disappearance, she maintained an upbeat demeanor for the reporter, who described her as having an "exuberant" personality. Beverly's real motivation for speaking to the reporter was to make sure the public knew of the $1,000 reward for finding the plane. The story prominently mentioned the reward.

Some of the most creative ideas for publicity came from the Rev. Allbritton, Sara's minister. In a Hollywood-type scene, his church sent a plane over small towns in the search area and dropped leaflets about the missing plane and the reward money. Another big-thinking proposal went from the Reverend to the New Mexico Governor's Office. He requested the State Police set up roadblocks in prime hunting areas to pass out flyers. The Governor's Office did not respond to the request. Undaunted, Rev. Allbritton phoned Ed Whaley to see if Gulton Industries had connections with the Governor. There is no record that Gulton attempted to contact the Governor on behalf of Allbritton.

Beverly's in laws, Mr. and Mrs. John Horton were long-time residents of Dallas, Texas. He was a fireman and also had a paper hanging business. He was approaching the age that he began to think of retirement from the Dallas Fire Department, but that was several years away. When the Horton's received word about Jon Dale, they drove to Albuquerque and moved in with Beverly the day after the plane was reported missing. They vowed to remain until a definite clue appeared. At first, that meant keeping Beverly company, attending meetings at Gulton Industries, participating in the ground searches, distributing the reward posters.

John and Vivian Horton returned to Texas for a short time in late October. While there, she contacted Democratic Texas State Senator Jack Hightower and asked if he could find out if there were any government reconnaissance missions in the State of New Mexico? If there were, could they possibly fly over the Pecos Wilderness? Jack Hightower was a Texas good ole boy, given credit for making comments such as: "The only things you find in the middle of the road in Texas are yellow stripes and dead armadillos." He didn't believe in waiting, he would try to find some answers.

Above all things, there was nothing more important to the Horton parents than helping in any way they could to find their son. But searching was costly. Upon hearing about the difficult situation of their friend and colleague, the men and women of the Dallas Fire Department held fund raisers to help fund the search effort. Combined with his income, Mr. Horton had enough money to contribute substantially to the search.

On the last day of October, John Horton came to visit Ed Whaley to review efforts. He told Whaley that he had over $1,000 and wanted help in deciding how to spend it in the most effective manner. Whaley told him that air search and photo work were the best ways. It was decided the money from the Hortons and from the fire department would go mainly to pay for helicopter flying time.

Beyond raising money, the Horton parents were busy contacting government officials to get more support. Little did the Hortons know that a telegram earlier sent to Washington, DC would bring more help.

President Lyndon B. Johnson sat in the Oval Office at the White House, inspecting the document in front of him. Reading the document undoubtedly gave him a good feeling. This is the kind of thing he was fond of—helping out our ordinary citizens. Johnson was credited for designing the "Great Society" legislation that included laws that upheld civil rights, Public Broadcasting, Medicare, Medicaid, environmental protection, aid to education, and his "War on Poverty."

It was great that the Air Force thought this proposal wouldn't endanger our national security. At the same time, useful training of the military could come from the mission.

Goodness knows, President Johnson needed a break from the relentless string of bad news that crossed his desk. It had been seven months since he shocked the world on March 31, 1968 and announced "I shall not seek, and I will not accept, the nomination of my party for another term as your president." From the pinnacle of respect reached in 1964 when he ushered and signed into law the Civil Rights Act, President Johnson was now loathed by much of the public. He had risked all his political capital on a major escalation of the Vietnam War, and the war was going badly. The bear of a man was now haggard and beat down, and he decided to return home to Texas.

After studying the document, he directed his pen to the paper. His signature allowed use of a key national strategic military asset—the U-2 spy plane—to be used in a civilian application in New Mexico. The Air Force started detailed planning for a photo reconnaissance of the Pecos Wilderness using the heralded Lockheed U-2 spy plane. Presidential approval was required for this mission, as with all previous U-2 flyovers.

Beverly Horton was working around the house on October 28th when the door bell rang. The messenger handed her a telegram from Western Union. Apprehensively, she allowed herself to read the telegram:

MRS JON D HORTON
11409 HAINES RD ALBUQUERQUE NMEX

PRESIDENT JOHNSON HAS ASKED THAT I REPLY TO THE RECENT TELEGRAM SENT BY MR AND MRS JOHN L HORTON. I WAS UNABLE TO CONTACT YOU ON FRIDAY, TO INFORM YOU THAT THE REQUESTED PHOTO MISSION WILL BE FLOWN TODAY, OCTOBER 28, WEATHER CONDITIONS PERMITTING. ANY LEADS UNCOVERED WILL BE PASSED IMMEDIATELY TO THE RESCUE CENTER. ON BEHALF OF THE PRESIDENT, DEEPEST SYMPATHY DURING THIS MOST ANXIOUS TIME.
OSAF. COLONEL B M ETTENSON, WHITE HOUSE
LIAISON OFFICER OFFICE OF THE SECRETARY OF THE AIR FORCE
WASHINGTON DC 20330

Although Mr. and Mrs. Horton's telegram was specifically referenced by the White House, it is possible that some additional inquiry by Jack Hightower may have contributed to the decision to fly the U-2. Hightower was a delegate at the 1968 Democratic National Convention and may have had opportunity to bring the subject up with his fellow Texan Lyndon B. Johnson.

On the morning of October 28th, a U-2 aircraft took off from Davis Monthan Air Force Base in Tucson, Arizona. The aircraft carried a single pilot wearing a fully pressurized flight suit. This was unlike few other aircraft on earth in appearance, as well as in its remarkable high-flying and long-range performance. The aircraft's glider-like wings enabled the plane to fly above 70,000 feet—higher than any other aircraft had flown—an altitude selected by designers to be above known radar and missile capabilities, and take high-resolution pictures. With its ability to loiter at high altitudes for six or more hours, the aircraft was a workhorse reconnaissance tool for the Air Force and CIA. There were only about two dozen of the aircraft in existence at the time.

That such an aircraft was being used in the search for the hunters was in itself a remarkable development. The aircraft's mere existence was a national secret until May 1, 1960 when, during an overflight of the Soviet Union that was personally authorized by President Eisenhower, a U-2 flown by Francis Gary "Frank" Powers was shot down by a surface-to-air-missile. After Soviet leader Nikita Khrushchev announced that an American spy plane had been shot down over central Russia, Eisenhower eventually confirmed that he had ordered the U-2 flights over a three-year period in order to protect the United States from a surprise attack. This event introduced the world to the U-2 aircraft. The value of the U-2 was demonstrated in 1962 when these aircraft made early discovery of Soviet attempts to install ballistic missile sites in Cuba. The U-2 is still used by the military.

Despite being equipped with state-of-the-art surveillance and photographic capabilities, the U-2 was one of the most difficult planes to fly. Its light weight gave the pilot little room for error. At high altitudes, the plane must fly very near its maximum speed to maintain altitude. However, the plane's stall speed at that altitude is only 10 knots less than its maximum speed. At lower altitudes, the operational control systems become physically difficult to manipulate (the mechanical flight-control system lacks hydraulic assist). The aircraft's high lift design is so effective it prevents a normal landing; instead, the pilot must stall the wing to touch down—think of it as a controlled crash landing. Because of the pilot's position in the cockpit and because the pilot's vision is compromised by the cumbersome pressure suit required at high altitudes, he cannot see the landing and must be talked down by a chase vehicle. Balancing is so critical on the U-2 that the camera had to use a split film, with reels on one side feeding forward while those on the other side feed backward, thus maintaining a balanced weight distribution through the whole flight.

In an attempt to optimize resolution of the pictures of the Sangre de Cristo Range, the pilot chose to fly the U-2 at a relatively low altitude of 28,000 feet, some 20,000 feet above the average terrain. At this altitude, the weather patterns that develop over the mountain ranges were added problems that had to be considered by the pilot.

It took three days for the U-2 to complete its flights. The film reels were immediately transferred to the Strategic Air Command, who in turn relayed the film to March Air Force Base in California for photo interpretation. Stretched out, the exposed film extended nearly a mile in length.

Whatever images the film contained were analyzed in great detail by the photo analysts. The analysts knew ahead of time that their job would be difficult because there had been a small snow storm three days before the U-2 flight. The white plane could effectively disappear beneath the recent snow. They hoped the images would be high quality; there was no chance to repeat the flights soon. A major snow storm hit the area November 1 and dropped up to three feet of additional snow in the Wilderness.

Even with their extensive connection to the defense industry, it took several days of inquiry for the Search Team to verify that the photo mission referred to in the White House telegram had actually happened. Eventually they connected with a Colonel Crane at March Air Force Base who confirmed the mission, and provided some details of the fly over.

Pictures were taken over a large swath of land 60 miles wide and 160 miles long, stretching from near Albuquerque north to the Colorado border. This translates to nearly 9,300 square miles of landscape to assess—an area larger than Connecticut—and included much of the Sangre de Cristo Mountain Range.

Film obtained from the U-2 aircraft flights arrived at March Air Force Base and was readied for photo interpretation. Being part of the Strategic Air Command, processing the film quickly was a goal. To assist the photo analyst in identifying unusual features in the film, the military had attempted to automate as much of the photo interpretation as possible. A large roll of film was loaded into an optical reader. As the images scrolled through the reader, the system tracked changes in light density. A major change in density would trigger a stop in the reader and the photo analyst could then manually inspect the image. The analysts discovered in no time that this was going to be a challenging batch of film to interpret. The first difficulty was posed by the forest itself: Alternating patches of afternoon sunlight and shadows in the trees caused the reader to stop often. A second problem was caused by the spotty cover of snow on the ground. Normally, small features like shadows and snow patches might be ignored by the system or the analyst. Here, such features have to be looked at carefully because they could be hiding a small airplane.

As the photo interpretation dragged on, the Search Team at Gulton provided a set of coordinates of old plane crash sites to the Air Force to help the analysts assess the requirements for finding a crash site. Also, they provided coordinates of Dalton Canyon and other major locations of interest.

16

It was a chance encounter. While attending Sunday services October 27th at a different church, Sara ran into an acquaintance who used to attend Sara's regular church. In the course of updating her friend on the search efforts, Sara's exasperation came through. It was clear that Sara would cling to anything that would help find Ken. Her friend listened intently and finally offered to put her in contact with a famous psychic, Dr. Gilbert Holloway. Sara had not heard of Dr. Holloway, nor was convinced of relying on ESP for help, but reluctantly agreed to speak with him out of desperation.

The next day, Sara Brittain spoke with Dr. Holloway who was working in New York City. It was the same day that Beverly Horton received the telegram from the White House. After some brief introductions, Dr. Holloway asked Sara a series of questions about her husband and the flight.

Considering what he had heard, Dr. Holloway buoyed Sara's spirits immensely by saying indications were that all or some of the men survived the landing. He also said that people in the plane should be found within two to two and a half weeks. It had now been nearly four weeks since the plane's disappearance. He requested that Sara send a copy of her husband's handwriting, plus some maps of the area. These items will help him narrow down the location of the plane.

Of all the famed psychics of the time, it was Dr. Gilbert Holloway that touched the common guy on the street. Unlike Jeane Dixon, who benefited from national media coverage, Dr. Holloway came to be known through over 100 appearances on regional radio and TV shows. He and his wife drove their mobile home cross country appearing on radio stations, earning money by doing private psychic readings after the broadcasts: WMEX in King of Prussia, Pennsylvania; WPRO in Rhode Island; WJAS and WDKA in Pittsburgh; KILT in Houston; KOB in Albuquerque, and so on.

His reputation grew through "listener participation" programs, particularly those on radio. All that was required was for a caller to the radio station to say in a clear voice, "Dr. Holloway, will you please tell me something about myself." Holloway would then share what he felt or sensed about the person's background and nature. In his book *ESP and Your Super-Conscious*, Dr. Holloway described what happens after the caller asks the question:

"With my eyes closed and listening intently to the incoming voice—which is my only point of contact with the subject—I receive from the Super-Conscious mind a *stream of impressions* relating to the person. These inner impressions are rapidly verbalized. They are given out over the air and this constitutes the ESP material of the test of demonstration....The wonder is that any of the information at all is relevant or evidential. The truth, confirmed by experience, is that there is a high degree or percentage of evidential, accurate personal information. The radio announcer likes to ask the caller, 'What percentage of accuracy do you give Dr. Holloway?' Usually they will say, 'About eighty-five percent.' Sometimes it is all true, one hundred percent. Sometimes it will be lower, but the average will run eight percent to ninety percent in most programs."

As implausible as this over-the-phone projection may seem, he *connected* with the audience. Wherever he appeared, it was not surprising for the radio station to receive a record number of calls during his appearance. For instance, in summer 1966 Boston radio host Jerry Williams invited Dr. Holloway to appear on his late-night program at WBBM, from eight p.m. to nine p.m. By the time the hour elapsed, the phone system had recorded over 29,000 calls attempted to the station. As further evidence of his popularity, radio station WMEX in Boston took the unusual step to alter their normal broadcast format to have Dr. Holloway demonstrate ESP for three hours each Tuesday and Thursday afternoon for a full year.

Dr. Holloway was simply mesmerizing to listen to. He was blessed with a clear baritone voice that was perfectly suited for radio. For television, his handsome looks helped—resembling actor Andy Garcia, or maybe Ed Murrow the stalwart CBS news anchor. He was brilliant and well spoken: He had studied at Stanford and Columbia Universities and held Doctorates of Divinity and Philosophy.

On the last day of October, 1968, Dr. Gilbert Holloway phoned Sara Brittain. His initial indications were that the plane was near Wheeler Peak—highest peak in New Mexico at 13,161 feet—between the small towns of Valdez (west of the peak) and Eagle Nest (east of the peak). He told her that some or all of the people survived the crash, and the people should be found within the next two to two and a half weeks.

Sara passed the information immediately on to Ed Whaley. After a brief follow up conversation with Dr. Holloway's wife, Whaley contacted the search team. Soon, small fixed wing aircraft focused on the area. Among the volunteer pilots flying there was Bob Rummage, a long-time friend of Sara and Ken Brittain. Rummage had flown in from Prescott, Arizona two days earlier, after Sara phoned asking for help.

The guidance from Dr. Holloway gave Sara energy. The next day she and Beverly flew with Bob Rummage to Taos, the largest town near Wheeler Peak, and left notices at grade schools. They barely beat a snow storm into Albuquerque on their return, and then Sara went to a late meeting with the New Mexico Mounted Patrol. The Patrol said they would send men into the Eagle Nest area.

Three days after his initial reading on the plane, Dr. Holloway had more insight to offer. Ed Whaley interviewed him on November 2nd.

"I am looking at Wheeler Peak, and this of course is not a detailed map at all," Dr. Holloway mentions. "I am seeing a road that is almost like a circle around Wheeler Peak. Looks like it may go around 30 miles or so. Eagle Nest is one part, Arroyo Hondo, and so on. The plane is down within about a 30 to 40 mile circle (15 to 20 mile radius of Wheeler Peak), something like that. It's that area—of course, it's my impression in the spirit. Most likely the southern half—the lower semicircle. I am looking at the map and see Palo Flechado Pass."

"This is the road that comes from Taos going over towards the town of Angel Fire and on up towards Eagle Nest. It crosses a pass—that's right," Whaley clarified for Holloway.

The Doctor continued. "The plane is north of that pass. I have a feeling that the plane flew over this, you see, flew over this pass and then as I go beyond there, I have the impression that the motor gave out. Either the motor gave out or the fuel line or something like that. I heard the motor fluttering, you know? I believe that their engine trouble developed in that area. Therefore, they went down somewhere from there (Palo Flechado Pass) and Wheeler Peak."

At this point, Ed Whaley attempts to share some logic, assessing what Dr. Holloway has said. Holloway's target area is north of the current search focus area, by 10 to 15 minutes flying time.

"Now this is the thing that is very peculiar to us and the reason that we have started our search south of this line (Palo Flechado Pass). You do realize that the southern extremities that you talk about borders on the northern area of where they wanted to go? The pilot said he was going to be airborne two hours, and he was only carrying 3½ hours of fuel. Just to fly from Albuquerque to that area (Palo Flechado Pass) is over an hour. So if he just wanted to go up there and turn around and come back and not do any ground searching—and of course they were looking for elk—they would be taking up their gasoline.

"This is the reason we have concentrated our search generally south of the area you are talking about. That doesn't mean that the pilot can't change his mind—see something in the area and go up there—and lose track of the gasoline and run out or anything. So the area you mentioned has not been given as detailed a look as down south."

"But the fact remains that you haven't found anything down south," Dr. Holloway responded.

"That is correct."

"You understand I am not an expert in this at all. I have done no private flying at all, but I'm just working this through the gift of the spirit."

"I understand what you are saying," Whaley said, trying to keep the conversation going. "I was just trying to give you more information."

Holloway persisted to press his opinion, "But it is possible that they could have gotten up there?"

"They could have flown up there, there is no question about it.... I was wondering if you had any feel for what time they went down? Because that would give us an indication and tie back into their gasoline and so forth."

"I don't have any particular intuitive feeling about the time. Mrs. Brittain called this morning and asked for direction," Holloway added. "I said they were going towards the north or northwest as they descended."

Whaley next offered to send Dr. Holloway a set of detailed aero topographic maps, in case he could refine his crash location further. Finally, the conversation turned to likelihood that the men in the plane are alive.

Dr. Holloway added, "You know the Bible speaks of the gift of miracles, and we are in the age of miracles. It would be a miracle if any of these men are alive but it is in that sense...."

"The way we feel about it right now is that it is one in a million chances and there is a possibility until someone brings a dead body, that the men could still be alive."

"It's remote, it would be a miracle, but there is that miraculous possibility. And, of course, you can't really rest on it, can you?" Holloway stressed.

That comment captured the drive behind the search—the need to find an answer. What happened to the plane and the four men?

Perhaps talking to himself as well as attempting to educate Dr. Holloway, Ed Whaley then spoke more about the realities of the situation.

"One of the big problems, to give you some insight, is that this land is all at very high altitudes. The lowest altitude you are talking about is probably around 8,000 feet and the high mountains go up to 13,161 feet. The temperatures in the night range from 20 degrees F to zero degrees F and so forth. If these people were hurt as you indicated the other day, it is very doubtful that any of them survived more than one or two nights. They only had on street clothes and unless they were able to get out and build a fire... And with all of them experienced hunters and mountain people, well I feel they could have walked out. They did have detailed maps. Between myself and you, I do not hold much hope that they could be alive."

Dr. Holloway still tried to maintain some optimism, "They could have walked to somewhere, couldn't they?"

"Yes. But, you see they have been down almost four and a half weeks now."

"And it is just as logical to think of them not being alive. I believe that too. There are so many problems there," the doctor concluded.

Two days later, Dr. Holloway called Sara Brittain again with a refined projection of the plane's location. The new location came after he had a chance to study detailed aero topographic maps sent by Ed Whaley. Holloway reported that, as best as he could tell, the plane was down between (or around) Lew Wallace Peak and Star Lake. This area is just southeast of Wheeler Peak.

TAOS INDIAN PUEBLO, NEW MEXICO

A major piece of land within Dr. Holloway's Wheeler Peak search area belonged to the Taos Indian Pueblo. Much of the higher elevations was considered to be sacred tribal lands, and visited only by selected tribal members. Consequently there was little hope of Whaley's team being given permission to enter the area and conduct a ground search. So, the team turned to the Taos Pueblo Governor and Lieutenant Governor for help. They asked Jim Kohl to contact the Governor. Jim Kohl was best man at Ken Brittain's wedding, and had been one of the mainstays of the search effort since the plane disappeared. After several attempts, on November 3rd he succeeded in meeting with the Pueblo's Governor and Lieutenant Governor, asking for help searching the tribal lands.

It is an extreme honor being voted Governor by the people of an Indian Tribe. At the same time, there is tremendous responsibility assumed with the position. Decisions must be reached depending on whether the action benefited the tribe as a whole—not the individual—and whether the action was consistent with tribal customs.

The Gulton team started the meeting by summarizing the search efforts conducted so far. They then discussed why it was necessary to search on Taos Pueblo lands. They ended their initial remarks by saying that a $500 reward was offered to the person finding the plane, thinking that the reward money would help encourage the tribe's participation. Instead, the comment had the opposite effect.

It was immediately evident that the Governor was troubled than *an individual* with the tribe would get the reward money, rather than the tribe itself. His concern was great enough that it appeared he wouldn't approve the tribe's participation in the search. But then he paused and asked, as a seasoned statesman might, something like, "Well, if we find it—*We the Tribe*—would you donate the five hundred dollars to the tribe?" When the Gulton team quickly agreed, the tribal leaders were convinced: They expected to put people in the field within two days.

It had been about one week since Jeane Dixon received a packet of information about the missing plane. Sara Brittain and Beverly Horton gathered on November 3 at the Motts' apartment in Albuquerque and awaited a phone call from the psychic. The waiting was made easier as they watched from the Motts' balcony an air show that featured a performance of the famed flying "Thunderbirds". Jeane Dixon eventually called from Washington DC. The psychic said that she had no "Vibrations" on this, and so far she has not used the material sent to her.

In order for her to become better tuned to the situation, she asked the wives to enter a "Novena" (a Catholic term), that involved a nine-day prayer and devotional. They were instructed to awake precisely at six forty-five a.m. Washington, DC time and spend 20 to 30 minutes in prayer and thought, with focus on their husbands.

Jeane Dixon called four days later and relayed a message through Mrs. Motts in Albuquerque. She reported that she knows what happened, but wants to meditate more.

Eventually Jeane Dixon phones Beverly Horton and Sara Brittain with her opinion. She said that the men were not looking at elk when it went down. Instead they were looking at some kind of a "wooly-fur animal," and crashed when they got too low. Later, she relayed that the plane is not down in mountains or canyons; the plane is down in a valley. No other opinions on the location of the plane or on the fate of the passengers were offered. All told, it was a sketchy reading by the famed psychic, and of no use.

17

As the last remaining leaves dropped from the Aspen trees early in November, the Pecos Wilderness settled in for a long, deep winter. Within days, the real heavy snows and blizzards would arrive and completely cover any remaining bare spots on the ground. It would be May before the cascade of snowmelt begins and the dormant yellow wild lilies that populate the high country would awake. Many north-facing slopes would remain snow covered until mid-summer.

No group was more keenly aware of what was coming to the mountains than the Search Team. This awareness prompted a phone call from Richard McVay of EG&G to the Gulton Offices. Could they meet to review what had been done so far and discuss what might be done during the coming winter months? On November 6, many of the key players in the search met in the Gulton war room.

As the 19 men entered the room that evening, they were greeted by Sara Brittain and Beverly Horton. The wives wanted all in the room to know how deeply appreciative they were of the team's continuing efforts to find the plane and men. In attendance were more than a dozen men from EG&G and Gulton Industries who formed the brains of the ground search team. As always, Rev. C. E. Allbritton represented the Eastern Hills Baptist Church. Nick Beers, the missing pilot's principal flight instructor was there, as were representatives from the Woodmen of America. There were two men from the Civil Air Patrol.

It was an especially heartwarming moment when the wives finally got to meet Earl Livingston, public information officer for the CAP. Up till now, they only knew his voice on the telephone. Every day without fail, he called the wives around dinner time with updates on the CAP search. He had become part of their families.

To make sure all the parties tracked together, the meeting started with a review of what was known about where the men wanted to hunt. Nothing new here: there still was a need to consider two distinct areas of the Pecos Wilderness that were separated by about 25 miles. The northern edge of the Wilderness near Tres Ritos was the intended hunting area of the Gulton employees, Ken Brittain and Jon Dale Horton. The southern edge of the Wilderness near Cowles was the second area of interest and the intended hunting area of Ron Jones, an EG&G employee. The only added detail that came up was provided by Jack Frost of EG&G. The present thinking was that Ron Jones planned to start at the top of the Santa Fe ski basin, and then hunt east across the mountains.

In preparation for the meeting, Frank Young and his colleagues at Sandia Corporation did a technical analysis of whether it was feasible for the plane to climb out of Dalton Canyon. To do the analysis, they combined data from the plane's manufacturer with basic trigonometry and geography.

Based on information from the Piper Cherokee manufacturer, they determined the aircraft achieved its best rate of climb at a speed of 85 miles per hour. However, the rate of climb is not constant and the aircraft climbs at a slower rate the higher it goes in elevation. The sightings by Stanley Gonzales and Georgia Smith placed the aircraft at an elevation of about 7500 feet, flying in a northeast direction. Starting at that initial position, the analysts calculated that under ideal conditions (no downdrafts and maintaining a speed of 85 mph), the plane would have flown horizontally along the ground a distance of 8,503 feet as it climbed an additional 500 vertical feet to an elevation of 8,000 feet. They repeated the calculations for other steps in elevation: 8,000 to 8,500 feet, 8,500 to 9,000 feet, and so forth. Finally, they laid these results out on a topographic map to see if the plane could clear the nearby hills as it tried to climb out of the canyon. They concluded that the Piper Cherokee might have made it out of Dalton Canyon, but only if the aircraft was functioning optimally and they encountered no downdrafts.

This analysis prompted a review by the Search Team of just how well Dalton Canyon was covered. The opinions were divided, but that was in itself enough to look further in Dalton Canyon. Searches in Dalton Canyon still had the highest probability of detection. There remained the narrow possibility that the aircraft made it out of the canyon. Also, the Search Team had to always consider that maybe the sightings were not reliable.

It was decided that one more ground search would be attempted during the coming weekend. This would be the final large ground search until next spring. EG&G and Woodmen of the World took the lead.

The following day, Sara Brittain drove to Taos Pueblo to talk with the Governor on the progress of the search and leave detailed topographic maps of Lew Wallace Peak, Star Lake, and Blue Lake. She also inquired if it was possible for the Pueblo to put additional searchers out looking; the extra searchers would include Indians from Taos Pueblo and non-Indians from Albuquerque.

A little more than a week after the U-2 flights, Col. Crane at March AFB reported on November 7th that no sign of the missing plane had been found. The analysts so far had spent 128 man hours of time and been over the 4,500 feet of film four times, and were still looking. They had not found the old plane crash sites yet because of the shadows. Colonel Crane felt that the pictures were taken at too high an altitude and Davis Monthan AFB should re-fly the mission over *specific* locations at as low an altitude as possible (5,000 feet lower would be ideal). He would ask the Air Force to support a flight at low altitudes over Dalton and Star Lake areas. There is no evidence that a subsequent U-2 flight was flown.

The disappointing results of the U-2 flights, with state-of-the-art equipment, underscored how difficult it would be to spot the missing plane.

As Beverly Horton walked through the tree-lined campus of the University of New Mexico one day in late October, she marveled how peaceful the place was, in contrast to her life. Desperate to try anything for answers, Beverly came to visit the UNM Library in attempt to contact Peter Hurkos, a famed psychic whom she had read much about. She looked through the Library's collection of California phone books, but she did not find any way to get in touch with him. In course, however, she came across information about a Maxine Bell, a Los Angeles medium that, Beverly came to conclude, was somehow involved with an ESP exploration unit at UCLA. After Beverly reached her by phone, the medium asked to be sent personal items that Jon Dale had used the morning of the flight. Beverly sent his razor, tooth brush, and a paper napkin he used for breakfast.

Soon, Beverly received a response from Maxine Bell. The typed-written cover letter explained that the medium was forwarding a tape recording made by a developing student of hers. The medium provided the caveat that "I lay no claim to her accuracy whatsoever as this was her first attempt 'talking in' to clothes and articles." Nonetheless, the medium felt that Mrs. Horton would benefit from hearing the tape.

On whether her husband was alive: The student and Maxine Bell apparently differed in opinion. "She (the student) feels that your husband is still alive but that the others are dead." Maxine stated the opposite: "Mr. Horton is dead. He died peacefully and not in pain. It was quick."

On the (alternative) reason for the flight: "Listen to the tape carefully. She saw them going north and east of Albuquerque. And she said your husband did not know what the others had in mind when they took off for he thought they were just going elk hunting or sighting for elk, but that the others had something more secretive in mind and didn't want to tell anyone what that was until they succeeded at finding what they hoped to find, some kind of a clue, etc.!"

On the cause of the crash: "I feel there was some magnetic pull on the plane from a hidden radar (underground) site which caused the plane to act up."

On Jon Dale Horton: "...Mr. Horton is very close to you now and is trying to talk directly to you. He is very disturbed that this may affect your condition and he doesn't want you to let it ruin a normal pregnancy. He is terribly concerned over you. He says he will be with you until the baby comes. He is just sick they took the trip. He has promised to communicate thru me whenever you write your letters and ask questions of him in the future months ahead.... I am a spiritualist by religion and know that there is no death of the personality even though the body is gone and that communication is possible between these two dimensions of life and so-called death. I am sure he will provide this to you as time goes on these next few months."

On the Medium's next steps: "The next thing I would like to do is his horoscope, in all seriousness, for his death chart will show the conditions and location of his death. For this I require, the day, month, year, city and exact hour and minute of his birth. Send me also the approximate time when he took off and how long he could have been in the air."

On finding the plane: "I feel that air planes would not be able to see where the plane landed because it is in a crevice and on its right side and can only be found by travel on foot or by donkey. I do feel that within a month there will be found the remains scattered about and what was left of the plane."

On communicating with her husband: "Again I am urged by your husband as I write you...to not grieve anymore as he is very able to communicate with you on out and that it no longer matters what happened to his body or the plane. To communicate with you does matter to him. So next letter enclose the full birth data...and questions you would like to ask him if you were actually here. We are in research...don't forget. The proof of the accuracy is in what will come thru from him. Willing to try?"

As Beverly listened to the tape recording, she latched onto a brief comment made about Jon Dale's location: *He is in a spiritual place—there may be even a church or cemetery in the area.* This, of course, piqued Beverly's interest, and she mailed a topographic map of the area to the medium, hoping more detail could be provided; there are many places in the Pecos Wilderness that have been considered to be a spiritual. In early November, Maxine Bell returned the topographic map with a mark on Penitente Peak. Beverly studied the map and noticed that the Peak was close to Holy Ghost Canyon and Spirit Lake. Could this be the spiritual place mentioned on the tape?

Beverly contacted the Forest Service to see whether they had anybody working near Penitente Peak. She was in luck. A crew was working in the area, installing small dams in the stream channels on the mountain. The Forest Service reported finding nothing unusual on the Peak.

Maxine Bell later asked for travel funds to come to New Mexico for follow up work. Beverly was unable to help, being short on money at the time, and the medium did no more readings. Because the Forest Service did not find anything, Beverly dropped Penitente Peak as a focal point.

18

Every time Ed Whaley tried to narrow down the search area, he came back to the timing of the sightings near Dalton Canyon. He sketched out on paper what his understanding was of the intended flight plan. The plane departed Albuquerque at six forty a.m. for a two-hour flight; return was expected by eight forty a.m. Flying at 100 miles an hour, it would have taken about 80 minutes to fly to the Dalton area and return, under normal operating conditions. That meant that they only had 50 minutes to look around for elk in the general vicinity, from seven twenty to eight a.m.

They would have needed to depart the Dalton Canyon vicinity by around eight a.m. to make it back to Albuquerque by eight forty a.m. Assuming that Stanley Gonzales did see the plane at a time between eight and eight thirty a.m., it was reasonable to conclude the plane was on the way back. That timing agreed well with the approximate eight o'clock sighting by Georgia Smith. Thus, if the timing of the two sightings were in the ball park, the plane must be down between Dalton and Albuquerque. After considering the gas consumption picture, he concluded that the plane would have had ample fuel to return to Albuquerque from Dalton, unless the pilot had difficulty switching the fuel selection valve from one tank to the other. If the pilot indeed had trouble switching fuel tanks in flight, Whaley calculated that the plane would be within a 40 mile radius of Dalton Canyon.

All of these calculations, of course, were simply speculation. They nonetheless would be helpful in directing upcoming helicopter searches and the final ground search. All Whaley could do was use every tool at his disposal to give the searchers the highest probability of finding the missing aircraft.

It had taken Mr. John Horton two weeks to complete the arrangements to fly private helicopters. The flights would be paid for mostly by the donations of the men of the Dallas Fire Department. A copter was rented and then trailered to the mouth of Dalton Canyon, where a base camp and a refueling station was positioned. The first flight started at eight thirty in the morning of November 11th. They took off in a westerly direction up the middle of Dalton Canyon to the first major ridgeline and then returned along the right side of the canyon. They checked the three escape routes out of the canyon that would be logical ways for a fixed-wing aircraft to get out, if it was in trouble. Next, flying just above tree tops, they flew the nearby ridges and canyons before refueling. These were known possible hunting locations, or locations where possible sightings were made earlier by spotters in fixed wing aircraft.

After discussions with Mr. Horton, the helicopter headed to Holy Ghost Canyon—where Georgia Smith worked—and its sister-drainages. They crisscrossed the canyon in north-south traverses, rather than simply flying up and down the drainage. Eventually they made to the head of the canyon and passed over three alpine lakes that sit at the foot of Santa Fe Baldy and adjacent high peaks. After another refueling stop, the helicopter headed north toward Wheeler Peak. High winds turned them back just before the peak, but they still were able to cover substantial amount of ground up to and beyond Truchas Peak. The coverage went beyond Tres Ritos, the northernmost leg noted in the Piper Cherokee's flight plan. They returned to base by keeping about half mile of trees between them and timberline. All total, the helicopters flew 8½ hours in two days with no sighting of the missing plane. They thoroughly covered Dalton Canyon, Holy Ghost Canyon, and their feeder canyons.

After that effort, Mr. Horton had little money left to continue the helicopter searches. They concentrated their few remaining hours of flying time on the area between Albuquerque and Dalton Canyon. So went the last big hope for spotting the missing plane from the air—at least until next year.

During the next weekend, ground searchers were busy. One group of ground searchers focused on the smaller canyons north of Dalton Canyon; another group headed to the Star Lake area, where Dr. Holloway recommended. No signs of the missing plane were seen.

19

There was so much to do to get ready. Three weeks earlier, Barbara Jones had decided to return to her native Florida. It was time to pack the last box and actually make the move—all of which she did essentially on her own. Beyond the packing, there was the constant effort to keep up with the kids and stay strong during her pregnancy.

There was not a doubt in her mind that returning to Florida was the right thing to do—her family was there. Barbara was the first wife to think that Ron may indeed be dead. Her decision to leave for Florida was bolstered by what she considered to be a sign received.

Barbara was sitting on her covered porch at night and offered a prayer aloud that she needed a sign. She would take any kind of sign. She had waited for weeks since her husband's disappearance and still no sign. Then she said to herself, half jokingly, 'If you want me to leave, how about sending an airplane by?' Then, as if on cue, a huge low-flying military transport aircraft passed overhead and the roar filled the air. This was a very unusual occurrence because the normal airport flight lines are miles away. Even during normal conditions, Albuquerque doesn't often see such a large transport plane anywhere in the skies. Later, she sat down with Sara and Beverly to tell them the story of her prayer and the transport plane. She had decided that Ron was not coming back and there was nothing further she could do here. She would return home to Florida.

Beverly Horton and Sara Brittain came to visit Barbara to say goodbye and bring her up to date on the search. Just as important though, it was one last opportunity for these three different women to share their common bond and feelings. Each lady tried hard to be upbeat and not bring the others down. However, eventually it became clear that they all had recently had very low moments.

Beverly Horton woke each morning thinking of Jon Dale, and he was in her last thoughts before sleep. The pain of his disappearance was still strong. In addition, she had increasingly become worried for Jon Dale's parents, who were staying with her in Albuquerque. In vowing to stay until some solid news developed about their boy; they had completely severed themselves from their regular lives in Dallas. While she admired their determination and loved their help and companionship, she indirectly felt a burden for their condition. She believed that as long as she remained in Albuquerque, so would they. Being so close to the search effort was taking a toll on them.

The only way to return Jon Dale's parents to the lives was for her to leave town, Beverly decided.

Discouragement hit Sara Brittain more and more often during the month of November. She knew that every phase of the search had failed to yield any answers.

It was getting harder to stay optimistic. Increasingly, Sara thought, "What am I doing here?" Or, "How can I possibly raise my boys by myself?" Sara's November 14th entry in her diary reflected a low period: "Very low, depressed, etc. May have to leave without knowing something."

Often, though, it just took something small to keep Sara and Beverly going. For example, four days after that low period, Sara "took new heart" after she and Beverly came across some heroic survival stories. One article was about a pair of men that lived eight weeks in the woods, but were never found in time. A second story was about one who lived seven weeks in the Yukon. Another example of the wives' grasping at the smallest item for inspiration happened on November 20th. The student of the "medium" from Los Angeles called Beverly. She reported that the men were at Capulin National Monument near the town of Ratón along the Colorado border. The student also believed that something was wrong or secret.

Beverly passed this on to Sara. Rather than reacting negatively to the "Something is wrong" statement, Sara chose to focus on the possibility that the men would be found and wrote in her dairy "Like to believe that." The following day, the Brittains called from Phoenix with some disturbing news. Mrs. Brittain went to the Doctor and found out she has sugar diabetes and gangrene in a toe; the toe may have to be amputated. More bad news to add to the plate. Still, Sara's diary entry implausibly read, "We're optimistic now. Can hardly see how!"

As the wives suffered, so did the parents of the men. Until the men were found, the grieving process could not truly start for the parents or wives.

For the first time in over a month, the parents of pilot John Fishel returned on November 22nd to Albuquerque from their home in Nebraska. More elder and frail than the other parents, they chose to remain in Nebraska during the searches. Their visit to Albuquerque at this time, however, was not random. First, they realized that the search must dramatically slow down during the winter months. They wanted to personally thank the Search Team for its hard work and to hear first-hand about the search efforts so far. Second, Thanksgiving was just a week away now and they wanted to have a closer connection with their son's life.

The Fishel parents endured the wait for news in Nebraska. This distance though did not lessen the impact on the family. One of the pilot's sisters, Betty, described some of the family's suffering:

"This was a very difficult time for our family and all the others as you can imagine. My Dad did all he could to try to help find my brother John. And he, being my only brother, left my Mother never being the same! She pretty much declined mentally after this happened."

The Fischel parents joined Sara for dinner at her house and they shared stories. This dinner continued a string of engagements that Sara had recently had with the parents of the missing men. Ken Brittain's parents had stayed with Sara a week in their latest visit, returning to Phoenix five days before. Beverly and Jon Dale Horton's parents joined Sara for dinner, prior to their moving back to Dallas.

On November 24th, Beverly and the Hortons departed for Dallas. They were making the move with much regret, but all felt there was little more they could do now in Albuquerque. Plus, it was not easy for Beverly to move around now with the pregnancy. None of them wanted to go. None doubted he or she would return next spring to continue the hunt.

Beverly's dad left for Texas with a U-Haul and was followed by Beverly and the Hortons two days later. Beverly trailed the Hortons in their camper through New Mexico and into Texas, in her Ford Mustang packed to the gills. (She had sold the blue van.) After two days on the road, Beverly left them and drove toward Nemo, Texas and her parent's home. When she got out of the car, her brother rushed out, took her in his arms, and they cried together for a long time. Nothing was said.

This left Sara and her two young boys as the sole remaining family members in Albuquerque.

On Thanksgiving day, Sara awoke to a dream of a funeral, her first such dream. She thought to herself, "Hadn't I asked for a sign?" Sara and the boys spent the day with Jim Kohl and his wife, who brought a full Thanksgiving dinner to Sara's house.

20

Search efforts continued over the Thanksgiving weekend. Fifty volunteers participated in a ground search in the Star Lake area, where Dr. Holloway indicated the plane was. EG&G, Woodmen of the World, Gulton, the Eastern Hills Baptist Church, and the Boy Scouts were there. Ed Young flew his plane over a broad area stretching from the Rio Grande to the town of Pecos.

At this point, there wasn't much left that the Search Team could do until springtime. Still not wanting to stop outright, they chose to again place reward posters throughout the region. Gulton Industries people targeted Post Offices, public offices, stores, and other public places. Dave Quintana went to Las Vegas; Fletcher Wilson and J. Olonia visited Santa Fe; Frank Duran drove to Taos, Peñasco, and small towns north to Colorado; Ed Whaley went west to Los Alamos and other towns in the Jemez Mountains; Gerry Roybal hit the Santa Fe Airport; Dave Wirenius took flyers to three private airports in Albuquerque and then flew to airports along the Colorado border; John Wheatley went to a glider airport out of Mora; and George Friberg took 100 copies to Rev. Allbritton at the Eastern Hills Baptist Church for mounting on student bulletin boards at the University of New Mexico and Albuquerque High schools (hoping to get some student help next spring).

Rev. C.E. Allbritton and other members of the Eastern Hills Baptist Church began to wonder what might happen to search efforts during the winter months. There was no doubt they were in this for the long run, yet what could they do when the mountains were ladened with snow? "Snowmobiles!" someone volunteered. A quick inventory of church members showed numerous snowmobiles could be made available. So the group contacted Sara Brittain and Ed Whaley. The church members next prepared a letter to William Hurst of the U.S. Forest Service.

As Regional Director of the Forest Service's Southwest Region, William Hurst had responsibility over some 20 million acres of public lands in New Mexico and Arizona. He was busy in his job. The 1960s was a tumultuous time for the Forest Service. The agency was inundated with lawsuits filed over its policies toward a wide variety of issues such as timber harvesting practices and public access restrictions. Despite these pressures, he continually urged his supervisors, rangers, and guards to become active citizens in the community and to get to know the local citizens and their problems.

Upon returning to work after Thanksgiving, Hurst was going through his large stack of mail when he came upon a letter that dealt with those men in the missing airplane. The letter from a Mr. Clair Evans requested permission to enter the Pecos Wilderness on "over-the-snow" vehicles. Attached to the Evans letter was a separate note signed by Sara Brittain and Beverly Horton in which they requested Evans' help in organizing a snowmobile search because, "We do not want to have to wait a seven-year period." This plea touched William Hurst.

Keeping with his philosophy of being personally involved in the community, he decided that a normal reply was not appropriate. So Hurst drove to Sara Brittain's house later that day and spent several hours visiting. He explained that he couldn't give permission to operate snowmobiles in the Wilderness because of federal law. Then he showed stereo aerial photographs of the Wheeler Peak area made in 1965, to give Sara a 3-D perspective of the mountain. Finally, he offered his agency's help. This simple gesture by Hurst brought Sara to think that maybe "he'll be able to do something for us."

When Hurst returned to the office, he drafted two pieces of correspondence. The first went to supervisors and district rangers of the Carson and Santa Fe National Forests. He requested that his staff post the reward notices in conspicuous places in their communities and continue to help with the search however possible.

The second letter went to Mr. Evans. He officially denied the request for use of snowmobiles in the Pecos Wilderness. The Wilderness Act of 1964 prohibited the use of motorized vehicles in a Wilderness except under very restrictive situations, like emergencies when human life is endangered. However, he encouraged Mr. Evans to consider running snowmobiles into "the vast amount of Forest Service land open for their use. An organized search in some of these areas may be helpful in locating the lost plane and its occupants." Such would be the goal of upcoming searches during the winter months by Clair Evans and other members of the Eastern Hills Baptist church.

Mr. and Mrs. Horton had been back in Dallas for less than a week, yet they were already feeling guilty about leaving New Mexico. They were thinking about coming back to Albuquerque and so they phoned Ed Whaley on December 2nd to get his opinion. Whaley responded that it was snowing hard and the entire search area was thoroughly covered—he didn't see what good they could do by coming back now. Reluctantly, the Hortons remained in Dallas.

Perhaps because it was the start of the holiday season or maybe because Barbara and Beverly had finally departed, but Sara began to withdraw. On the Sunday after Thanksgiving she wrote in her diary, "Missed Kenny more than at any time since this happened. Looks like I will have to leave in January." That ended a string of 334 consecutive daily entries in her diary during 1968. There were only three entries made during the entire month of December, and no more entries thereafter until 1973. Her initial list of destinations to move to included places where she and Ken had close friends: Huntsville, Alabama; Houston, Texas; or maybe even her home town in Illinois.

Some good news came to Sara on December 5th. Much to her surprise, one of the life insurance companies notified her that they were sending her a check on Ken's presumed death. This would greatly stave off any pending concerns about finances. Other insurance companies chose to wait until their investigations of possible foul-play were concluded. Insurance adjusters were now interviewing the men's friends and co-workers at EG&G and Gulton Industries. Was there a chance they could have gone to Mexico or Cuba? How mentally stable was the pilot at the time of the trip? Were the men truly planning an actual hunting trip?

Meanwhile, Barbara Jones and Beverly Horton wrote letters to Sara once they were ensconced in their new homes. Barbara was living with her parents and would stay there at least through the birth of her third child, Terry. She believed that things eventually would become more stable and she could get her own apartment. Beverly had moved in with her sister in Forth Worth and, like Barbara, looked forward to delivering her child in the spring. However, there was little doubt that Beverly would return to New Mexico next summer to resume the search.

In one letter, Beverly told Sara that she had been to see a numerologist. According to Beverly, the numerologist did some work for a major petroleum company with respect to oil well locating. Her readings showed that the plane would be found in March 1969, the same month predicted by Dr. Holloway. This was a slightly positive forecast that the wives could hang on to. Still, any upbeat swing would be temporary. Sara's last diary entry for 1968 was on December 23rd: "Very bad day—harder to be here (Albuquerque) than at home (Illinois). Miss Kenny even more here."

Sara would be visited at Christmas by Ken's sister Babs and her boys. On Christmas Eve, 1968, they settled down and were transfixed watching the live television transmission from the crew of Apollo 8 orbiting the moon. This was ironic because Gulton Industries in Albuquerque had built the command and control modules for Kennedy Space Center from which Apollo 8 was launched. Astronauts William Anders, James Lovell and Frank Borman read the first 10 verses from the Book of Genesis in the Bible, to a rapt audience of an estimated two billion people back on Earth. NBC News anchor David Brinkley eloquently captured the importance of this moment as 1968 crept to an end:

"Nineteen sixty-eight was one of the most tumultuous years in the postwar era.... Numerous times in nineteen sixty-eight we have seen humans at their worst. By the end of sixty-eight we have seen the human race at its best, and at our best we are pretty good."

For the Brittain, Fishel, Horton, and Jones families, the year ended in sorrow and uncertainty. At the same time, however, they also witnessed humanity at its best, as countless friends and strangers volunteered what they could to help them find their missing loved ones.

21

1969

The normal bustle of the Holidays and visitors had passed and Sara Brittain spent time reflecting what her next step in life would be. Regardless of where she thought about moving, she felt unsettled. Finally, after much contemplation, she decided that she couldn't leave Albuquerque until she knew something concrete about Kenny's fate. She wouldn't feel right about starting a new chapter until the current chapter was resolved.

In theory, establishing a new home in a new location might direct her attention elsewhere and provide a distraction. Any such distraction, though, would be temporary. She believed that she would always regret not doing everything in her power *now* to find an answer. No answer today? We'll try again tomorrow.

Each wife dealt differently with the decision to stay or go. Each felt that their decision was the correct decision for their family at that point in time, and the others supported them. Having made the decision to stay, Sara began to think about the next steps. Letters and phone calls led Beverly Horton and Sara to decide to join forces during the coming summer and become a two-woman search team.

The mountains hadn't shown their secrets to any of the searchers so far. Owing to the accumulating snows, it would be many months before any hiking could be considered. This snow-related respite provided Sara with a break from search activities, and she returned her focus to regular daily events—raising her boys and playing the piano.

The engineers and technicians at Gulton Industries used the winter and spring months to get back to running their business. Only occasionally did search-related things cross Ed Whaley's desk. The leads still trickled in, though. Whaley dealt with the leads as they came. For instance, in the middle of January, John Wheatley reported seeing a shiny object as he flew his glider plane along the Rio Grande just north of Albuquerque; a follow up flight by Dave Wirenius found that the object was a mine shaft covered over with a plastic tarp. Another phone call to Whaley indicated that a helicopter had spotted a downed airplane north of Taos; Earl Livingston of the CAP eventually determined that it was a false report.

In mid-February, a colleague of Whaley reported that a man he works with talked to a woman friend who claimed to have seen a blue and white airplane near a small village just north of Santa Fe the morning of the disappearance. When the snow clears, the colleague will probably take a Jeep and investigate. (It is uncertain if the colleague did investigate.) Whaley later noted that several reports have been received that something was seen in this particular area. Perhaps it might have been possible for the plane to have used Dalton Canyon to cut over to the Santa Fe side of the mountains?

One hallway of the Gulton Industries building was lined with large-scale maps of the Pecos Wilderness area, where the plane was thought to be down. The maps were marked with routes followed by ground searchers and search aircraft. Massive files recorded various phases of the search.

Morys Hines, drafting supervisor at Gulton, sent out nearly 300 letters to Boy Scout leaders around the state, asking them to be alert for the plane during hikes into the area; Hines had been told that there are about 500 Boy Scouts in the northern mountains each weekend. Shortly thereafter, in response to that letter, Ed Whaley received a phone call from an Albuquerque scout master who planned to take a Boy Scout Troop into the Wilderness the end of May. The troop would start near the northern boundary near Truchas Peak, scale several 12,000 foot peaks as they headed south, and exit at the Santa Fe ski area. He offered to investigate any areas along the way that the Gulton planners desired.

Bill Carpenter and Ed Whaley continued to press the insurance companies to pay the wives their life insurance indemnities. Carpenter had argued more than once against the possibility that the men could have skipped to Mexico. Carpenter and Whaley were relieved when they heard that some of the insurance policies had been paid. In early January, the Gulton managers agreed that after three months it was time for the company to stop the regular salary payments of Ken Brittain and Jon Horton, now that the life insurance policies were being paid. The entire Gulton team had worked hard to ensure that the wives could pay their bills during Christmas and beyond.

Although business as usual had returned to Gulton, inquiries continually came to Whaley about the status of the missing men. The questions came from people within the Albuquerque office, from the corporate offices, from the news media, as well as from the insurance industry. The calls from the insurance industry, in particular, caused him to wonder if some clarification was needed. He prepared a statement:

February 12, 1969
AFFIDAVIT
Re: Jon Dale Horton and Kenneth Paul Brittain

To Whom It May Concern:

Please be advised that as of this date, the status of both of the above men is – "missing and presumed dead". Both Mr. Horton and Mr. Brittain were aboard a private aircraft which disappeared on October 2, 1968. Both men were employed at Gulton Industries and were under my supervision. I have been personally involved in ground and air searches conducted by the Civil Air Patrol and by private parties over the area of the Pecos Wilderness of New Mexico which was the area designated by their flight plan. Since no evidence has been found indicating the fate of the aircraft or its passengers, the subject personnel have not been declared legally dead.

A clipping of one of the many news items appearing in local newspapers is attached for further information.

Signed: E.W. Whaley
Engineering Manager

This letter also provided the wives a legal vehicle for filing income tax. The Government would not declare their husbands dead, so they were allowed to claim them as dependents. Beverly's taxes were audited in 1972 because of this. After reviewing the files, the auditors could not discredit her filing and ended up sending her an apology.

22

Beverly had an appointment with her doctors every week in March and April. Douglas was due April 1, 1969. He was late. On April 14th, she went in for the next check up and the doctors told her they were going to induce labor. She asked them if they could do it on the 17th? That date was the Hortons 30th wedding anniversary and she thought it would be a nice gift to them. Her parents joined the Hortons on the 17th to start the labor. They waited all day for Douglas to appear; it was later in the evening when she first got to see him and hold him. The next day she received 30 red roses.

As Beverly Horton held her newborn for the first time, she sighed lightly, "My beloved Douglas!" As promised, she had named the boy in honor of the stunt pilot Doug Rhinehart who had made such an impression on her last October. It could have been a day wrought with conflicting emotions, but the sheer joy of holding her nearly 10-pound baby was powerful and calming. This moment had been a long time coming. A jaundice perspective might say it actually somewhat surprising that it actually happened at all. For most couples, the maddening on again/off again seven-year courtship would have been more than enough to end the relationship. Indeed, she had repeatedly warned herself not to get involved with him again. Yet his good-natured charm always won her back—at least until the next break in the relationship. With time, however, the pair developed a strong bond of dependence on each other. On this day, the warmth she felt toward her son reminded her of some of Jon Dale's endearing qualities.

At work, Jon Dale was a meticulous engineer. His handwriting was as clear and precise as a draftsman. He excelled at designing prototypes of test equipment and carrying out experiments. Once he left work, though, an alternate personality came out that was relaxed and less concerned about the minutiae. He was now an affable "Aw, Shucks" character straight out of the old Andy Griffith TV series, according to a fellow co-worker.

"Sometimes, he just didn't seem to think things through," Beverly said with a smile. He had so much confidence is his innate ability to rectify any problems resulting from poor planning, he essentially chose to "wing it" when away from work.

No doubt it would have been thrilling for Beverly to share the birth of Douglas with Jon Dale. For now, however, she wanted to follow Jon Dale's lead and just enjoy the moment.

It is said that time speeds up as you age, but Jon Dale's parents felt that time just crept along during their winter stay in Dallas. According to what they had heard from Ed Whaley, the snows continued to come to the high country of New Mexico. In a letter, Whaley told the couple that in the first week of May the Sandia Mountains outside of Albuquerque received one foot of snow and "the northern mountains got at least the same amount, so I do not believe you can conduct an effective search until after June 1, 1969."

In the same letter, Whaley told the Hortons about the mailings to the Boy Scouts leaders. Mrs. Horton wrote back, thanking Whaley for Gulton's efforts, and said "Little boys have a way of getting around and into things more than adults and would be very likely prospects for finding the plane."

On May 26th, Jon Dale's parents finally left Dallas for New Mexico. After a frustrating absence of six months, they were returning to hunt for their son. They drove their camper and towed a newly-acquired jeep. The jeep was purchased to help them travel a web of rough dirt roads that occasionally penetrated the forest adjacent to the Pecos Wilderness. They were accompanied by a man and his son from Texas who brought two horses.

Like the Hortons, Sara Brittain got back to searching toward the end of May. In her first trip of the season, Sara searched for a week with Ken's father, who was visiting from Phoenix. Next, she headed back to Taos Pueblo to talk to the Governor and the War Chief. She convinced them to announce the plane's disappearance to *all* Pueblo members, and not just selected searchers.

The Hortons stayed at the Happy Valley campground in Holy Ghost Canyon. In short order, they were joined by Beverly and Sara, who slept in a small tent near their camper. Beverly's mom and Sara's niece from Illinois,

Pam Garrett, remained in Albuquerque to take care of Douglas and Sara's young boys while the wives were searching. Sara and Beverly would drive to Sara's Albuquerque house on the weekend and shower and be with their kids and parents, only to return to jeep or hike. Every day was a whirlwind of activity for the wives. Much of the time was taken up interviewing residents of small towns. Occasionally a lead would develop, which they would pass to Ed Whaley. Sometimes they would hike. They scrambled up many peaks to the snow line.

One weekend Sara sat with a reporter from the *Albuquerque Journal* which was preparing a long story about the missing men. The story appeared in the *Journal*'s Sunday edition on June 8th. Sara was quoted extensively in the article.

There's not much hope. But, the need to know makes the families keep looking, she told the *Journal* reporter.

"It's something that's not definite. I don't have any hope, but as long as the plane hasn't been found.... It's difficult to go ahead and make a complete adjustment until the plane is found. I think the other families feel that way too.... I believe the plane will be found in the next few weeks."

Sara based her optimism on a recent report from an unidentified person who said he saw smoke in the Wheeler Peak area the day the plane disappeared.

Almost by design, calls flooded into the Gulton offices after the *Journal* story. People called to report that they had seen something back in October 2nd, 1968, but they hadn't bother to report the incident. Several callers mentioned the Taos area. Why didn't they report this last year? They weren't sure what they had seen—perhaps the smoke was just a campfire. Tirelessly, the Gulton folks investigated each call: You never know which call will be the one to solve the mystery.

A new wave of volunteer searchers began to connect with Morys Hines and Whaley, mostly in response to the *Journal* story. There was the mining prospector who hiked throughout the Wilderness for one month starting in July. There were the Air Force Survival School participants who trained in the Wilderness. There was the Public Service Company of New Mexico employee who horse packed into the Wilderness three times each for four to five days.

Then there was Harry Tucker. Mr. Tucker was retired and spent most of the summer looking for the plane with his horse, a large mule, and two small female donkeys called Jennies. Mr. and Mrs. Horton paid for his personal food and expenses, and provided transportation for the animals to the search area.

The Jennies made it much easier to cover ground than by hiking and climbing. The Jennies were usually extremely sure footed, which allowed the riders to focus their attention on searching rather than walking. One day it started to rain while the wives were out. Sara put on her red rain parka and Beverly donned her blue one, and they rode back to camp on the jenny. Upon seeing them, Mr. Horton thought it was Mary and Joseph riding in.

On one ride through upper Dalton Canyon, Beverly spotted something shiny in the brambles just off the trail. She got off the jenny and struggled through the bushes, to find a pair of men's glasses. Amazingly, it was later discovered that they belonged to Ron Jones' dad, who had lost them the previous year during the first ground search.

Toward the end of June, John Fishel's parents arrived from Nebraska. They planned a two-week stay and offered to help where they could. About the same time, Sara got word that Barbara Jones might make a return trip from Florida later in the summer.

A couple in Albuquerque contacted Ed Whaley to notify him that they had used a spiritualism approach to finding the plane. Last fall they got a pencil from Jon Dale Horton's desk and saw visions of Jon. The man said that he knew that Jon was down near Elk Mountain; he marked the area on a topographic map for Whaley. On the couple's second trip into the area they found some old bones, which they gave to Whaley. Anthropologist friends of Whaley looked at the bones and determined they were from an animal and were not human.

23

Finding the missing Piper Cherokee was the narrow focus of so many people: family members, workers at Gulton Industries and EG&G, and fellow church members. However, when Ed Whaley opened his mail on July 1, 1969, he came to appreciate that the universe of missing aircraft was much larger than he had realized.

He was holding a rather bland-looking four-page federal government report entitled "NOTICES TO AIRMEN: OVERDUE AND MISSING CIVIL AIRCRAFT." The document stated "The following civil aircraft reported overdue and missing during the period 1 January 1968 through 30 April 1969 remain unlocated as of 15 May 1969." In total, the report listed 38 aircraft that had gone missing nationwide in that six-month time period. In the Central U.S. section, which included New Mexico, there were four listed along with *our* missing Piper Cherokee. Whaley quickly came to understand the magnitude of the jobs that Search and Rescue operations had. In retrospect, it was amazing that we got as much response from the federal government as we did, he thought.

The Eastern Hills Baptist Church was a major force behind the ground searches the previous year. Now that the high country was accessible again, the Eastern Hills Baptist Church decided to start the 1969 search season with another large organized ground search. Pleas for volunteers were placed in the Albuquerque newspapers. The 60-person group met Sara and Beverly for a two-day search of Dalton Canyon and tributaries.

To some people on the Search Team, it seemed as if Dalton Canyon had been thoroughly scrubbed and the Church was duplicating efforts. Ed Whaley and Mr. Horton, however, were not as confident. Whaley recently received information from the CAP that there were at least two known old aircraft crashes in Dalton Canyon. However, no searcher in the present effort had yet spotted these old crashes, indicating that the canyon could easily hold secrets.

Since the flood of calls early in June, the number of solid leads coming in to Gulton soon diminished. Apart from the recent Eastern Hills Baptist Church's ground search in Dalton Canyon, most of the searches now were being made by individuals or by small groups on weekends. Morys Hines, George Friberg, and Ed Whaley—and occasionally their wives—sometimes teamed up to explore an area. The pilots John Wheatley and Dave Wirenius looked from the skies when possible. Mr. Tucker continued to comb the backcountry with his mule and Jennies.

The one group of people which showed no signs of slowing down were the family members. They followed every possibility and jumped from one mountain peak to another. Mr. and Mrs. Horton spent half of the summer on the search until medical issues prevented them from staying longer. Although the odds of the families walking upon the plane were infinitesimally low, they continued to look whenever they had an opportunity.

Beverly and Sara hiked in a dozen different locations, including several scrambles on Wheeler Peak. One of the Wheeler Peak searches was directed toward a burned area, near Dr. Holloway's predicted crash location. Two local teachers from the village of Angel Fire joined the families on the burned area search. The teachers came by early in the morning to pick up Sara and Beverly's dad, Jim Payne, in their white Ford Bronco. Mr. and Mrs. Horton came in their Jeep. When they arrived at the burned area, they split up into pairs. They were going to cover as much ground as they could by walking far apart, while still keeping their partner in sight. Each pair was to walk for about an hour and return. Beverly's dad had heart problems, so he remained at the base with the vehicles.

Sara and Beverly paired up. The wives had walked for about 30 minutes climbing down the slope, always staying close enough to hear or see the other. They came to a steep drop off, and Beverly went to the right to climb down. Sara went left. By the time Beverly got to the bottom, she couldn't see Sara. Beverly called out, but Sara did not answer. Beverly moved more to the left, shouting for her, but she did not answer. Then to Beverly's right came the Hortons calling. They joined up to search for Sara. No Sara. They decided that when Sara lost contact, she probably would turn back to base. They returned to the base, but she wasn't there. Soon, the teachers returned and they all went back out to search for Sara. With darkness growing, they built a large fire at base and hoped for Sara's return. The teachers were preparing to drive to town to report her missing when Sara drove up in her car which she had left at the motel that morning. Sara told them that when she discovered she was lost, she didn't want to climb back up. She thought if she went downward she would at least come out at the highway. For hours, she descended the steep hillsides via a creek bottom until she walked into a hunter's camp as darkness approached. The hunters offered her food and shelter, but she refused, saying, "Just take me back to my people. I don't want them to worry." So the hunters drove Sara to her motel.

The group was back together now. It was dark and they had to drive down the mountain. They soon realized that the Horton's Jeep had no headlights—it had only been used before in the daytime. They descended using a caravan approach: the teachers drove the Bronco in front of the jeep, while Sara, Beverly, and Mr. Payne followed behind, illuminating the way for the jeep. They finally reached the highway at the base of the mountain.

Suddenly, they were surrounded by cars from the State Police and the local Sheriff. At first, they thought the officers were after the unlighted jeep. Beverly got out of the car to explain they were trying their best to drive safely, but was stopped by a shotgun pointed in her face and shouts for her to get down. She did. In short order, everything was cleared up. They were after the white Bronco. It seemed there had been a bank robbery that evening and the getaway car was a white Bronco. The group had had a rough day, but they laughed with relief as the law enforcement sped away in search of the real bank robbers.

24

During the second week in August, Ed Whaley composed a letter to Beverly Horton, who was home in Forth Worth. One purpose of the letter was to update Beverly on various search activities that she was interested in. The other purpose was to offer his opinion about the odds of finding the missing airplane.

He first wrote that the New Mexico State Police had investigated reports of smoke seen near Tres Ritos on the day the plane vanished and firmly concluded the smoke was "definitely smoke from burning brush." Another lead was pursued when a backpacker in the Wilderness reported that he was passed by a group of men on horses carrying what looked to be parts from an airplane. The State Police and Forest Service rangers tracked down the men and determined that they were legally salvaging parts from a two-year old crash near Hamilton Mesa; it wasn't the missing Piper Cherokee.

It was the third paragraph in the letter that carried the biggest impact. Whaley wrote as a friend:

"Beverly, I think it will be a long time before the plane is found. It must be located in an area that is very remote. I have told Sara and I repeat to you that from this point on I would recommend that you plan your life as if the plane will not be found."

As a member of the Eastern Hills Baptist Church, Coy Davis Jr., was good friends with Ken and Sara Brittain, and became passionately involved with the search. Although he participated in some of the ground searches, he found his niche through his writing skills. Many times he would compose letters asking for more resources from public officials such as U.S. Senator Joseph Montoya, Governor Dave Cargo, and Regional Forest Supervisor Hurst.

On August 19th, Coy Davis penned a letter sent both to Gulton Industries and EG&G. The letter sent to Gulton read as follows:

"Dear Sirs: I would like to inquire as to your willingness and request your company to post jointly with EG&G funds by the sum of $5,000 (each company donating $2,500) as a reward collectable until September 30, 1969. This will be widely publicized by the news media and will be payable to anyone who provides information leading to the discovery of a missing aircraft lost over northern New Mexico on October 2, 1968. Two of the four lost men were, I believe, employees of your firm.

"I stand to gain nothing personally from this request but am making it on behalf of the families of these missing men. This is not their idea and they do not know that I am making this request. I believe a $5,000 reward might bring results on the missing plane."

Upon reading the letter, Ed Whaley phoned Coy Davis. He learned that Davis got the idea from a missing plane incident in California where a $10,000 reward was posted, bringing out many additional searchers who found the plane. The reward stimulated people to look. Davis believed a larger reward would help stimulate people to look for the missing Piper Cherokee here, where $1000 would not.

Gulton Industries and EG&G declined Coy Davis' request. Each company felt they were participating in active search efforts, and their money was better spent in those efforts than in additional reward efforts. In the search, Gulton Industries alone had contributed many thousands of dollars of effort.

Coy Davis immediately responded to the companies. He thanked them for their consideration and past efforts, and then went on with an explanation of his beliefs on the matter:

"Nevertheless, I personally am quite convinced over the inability and the fruitlessness of the various searches that have been conducted to locate the missing airplane. I, personally, participated in some of these search efforts and I am convinced that it is a hopeless task to expect that this airplane will ever be found in any such organized search activity.

"As a personal friend of Mrs. Sara Brittain and her missing husband, Kenneth, I cannot simply find it within my own conscience to write this off as a hopeless endeavor. At this point I think the peace of mind of the remaining surviving wives is worth more than money can compensate for. If I had $5,000 or even $10,000 to personally invest, I would not be able to imagine a more worthwhile project.

"I am not asking you to reconsider your decision but I am simply giving you one person's viewpoint. I do not intend to give up on finding some way to help Mrs. Brittain and the other remaining wives recover their missing husbands if it is humanly possible to be done."

25

In the view of some guys, Stanley Gonzales lived a perfect and simple life. He owned a country bar and set his own hours. He regularly hunted in his backyard mountains. He didn't worry about time; in fact, he didn't carry a watch. After rising with the sun, he would feed his animals and then relax with a quick morning nap before the mailman arrived.

However, since he reported seeing an airplane fly low over Dalton Canyon the morning of October 2, 1968, his life had become more complicated. He now was visited often by a variety of people, mostly from Albuquerque, who wanted to hear the story of when he saw the airplane. A few weeks ago early in June, it was that gentleman John Horton. Although Stanley had talked with him several times last year, Mr. Horton wanted to talk again to see if Stanley remembered something different. The story did not change with each telling.

The latest visit was toward the end of June by a fellow named Bill Trembly. Once again, he told the man that he saw the plane take a sharp bank up into the canyon when it was flying only 20 to 30 feet above the trees, and described it as being a white or light grey plane. The man asked Stanley what time was it when he saw the plane. Stanley answered "About eight o'clock in the morning. The mailman hadn't come yet." He went on to say, "That plane was flying a little low for comfort for me."

When Stanley was not giving interviews or running his bar or hunting or napping, he looked for the plane himself. Once he rode along with the Las Vegas Mounted Patrol. But most of the time, he drove the back roads that went near the canyon, and he hiked. He usually looked in Dalton Canyon because, as he had said in many interviews, unless the pilot knew the lay of the land, he would have trouble getting out of the canyon.

It didn't take long before Stanley realized that there was a lot of ground to cover and the forest was very thick in places. He decided he needed help. He hired a local man to look for the plane.

Since his return to New Mexico, John Horton occasionally checked in with Ed Whaley to update him on his doings and findings. He typically was even-tempered about the search in these calls—objective and thinking about the next step. However, when he phoned Whaley on September 25th, there was an edge to his voice. Their conversation went something like this:

"I stopped by to visit with Stanley Gonzales."
"The one who spotted a plane in Dalton Canyon," Whaley said.
"Yes, that's the one. He told me he had hired a man at ten dollars a day to look for the plane. Well, it seems that on the fourth day out, the man found some burned and cut oak trees near the top of a cliff. But he couldn't get up to the area, so he came back and told Stanley."
"So does he want our help in getting to the cliff?"
"No. He told Stanley that he wanted five thousand dollars to lead people to the site. He thinks the burned trees are where the plane is."
"He wants money to show us the area. You're kidding."
"No. But wait until you hear the best part."
"This should be good."
"The man was shot in front of Stanley's bar a couple of days later before Stanley could do anything about this. The man was treated at Santa Fe and then taken to El Paso by his brother."
"Amazing."
"Stanley told me he wants me to give him five thousand for the man. But the man is not physically able to walk to the crash site and no one has actually seen the plane."
"You can't make this stuff up, can you?"

With hunting season soon arriving, many in the Search Team remained hopeful that hunters might come across the plane. So when Ed Whaley received a call from a hunter asking to help, he invited him to the office to look at Gulton's maps.

Whaley talked to the man and found the man had a desire to look for "our" plane. The man and three others would hunt the Pecos for elk and deer. Plus, one of the men in his party was a pilot and they might do some reconnaissance flying. This was exactly what Whaley wished for.

However, the man continued to tell Whaley that the *actual* reason for the visit was to tell Gulton about his cousin, who was not local, and her ability as an ESP type. (It would have been easy for Whaley to be amazed at how many ESP types offered to help in the search, but his notes from the meeting reflected no such reaction.) The man said that if he could have a personal photograph of the most aggressive person on the plane he would sent it to his cousin for her use with ESP. Whaley responded that they would try to comply. Then Whaley also showed him the maps and areas where other ESP types had said the plane was located.

John Fishel had two flight instructors. His regular instructor was Nick Beers, and they usually flew in the afternoon; Merl Malin was his teacher for the morning flights. Since the plane's disappearance, Nick Beers had taken to studying the files at Seven Bar Flying Service. Now, more than a half-year later, Beers was starting to question his initial judgment of John Fishel. In contrast to what he had told the press and the CAP when the plane disappeared, he was concluding that his student John Fishel was not a good pilot. He shared his concerns with Morys Hines at Gulton and with Sara Brittain. It is uncertain what motivated the sharing of his concerns.

There were several items he discovered that indicated to him that Fishel was careless, or perhaps even deceptive. To start, Beers found that Fishel had violated flight plan protocol by not designating the flight as a *cross country flight*. If a pilot is expecting to travel more than 20 miles away, the trip would be considered to be a cross country flight. With such a designation, Fishel's proposed flight would have been scrutinized more thoroughly by airport personnel, and perhaps denied.

Then there was the matter of John Fishel not being forthcoming about his flight. Beers went up with Fishel on October 1st (the day before the plane disappeared) and Fishel stated that he was going to the Pecos but did not say in what manner (plane, car, etc.), even though he clearly knew. Similarly, the daytime instructor, Merl Malin, ran into Fishel at Alameda Airport just before his departure for the Pecos Wilderness on October 2nd. Malin reported that Fischel did not mention anything about his pending flight. Beers also found that Fishel did not sign out that he was taking any passengers.

These actions indicated to the instructors that Fishel was trying to keep the flight secret from the very people who knew his flying abilities the best. Knowing that Seven Bar did not allow its rentals to be used for hunting purposes may have contributed to Fishel's silence. Combine these actions with the decision to fly a loaded aircraft into a risky environment (near mountains rising higher than the plane's service ceiling with downdrafts), along with the pilot's inexperience, and it seemed evident that his judgment was questionable.

One of the mysteries in this story is why the flight took place at all. From the information available, no concerns were ever raised by the men about the pilot's experience. In their professional lives, the engineers were paid to anticipate ways that equipment or electronics might fail, and then to design redundant fixes for these potential problems. Most importantly, the engineers knew they had responsibility for finding these problems. Thus, they always thought in this mode at work. When it came to the flight, however, the human dynamics were different. Ken Brittain didn't question the flight because he was invited at the last moment by his colleague and hunting partner, Jon Dale Horton. Jon Dale didn't question the flight because he trusted the judgment of Ron Jones, whom he knew well from their days at EG&G in Las Vegas. Ron Jones probably didn't question the flight because he worked with John Fishel on a daily basis, and found him to be reliable. Moreover, Fishel was a *licensed* pilot and Jones did not have the *credentials* to question the soundness of the flight. It wasn't the responsibility of the passengers to evaluate the risks; that responsibility lay solely with the pilot.

The only person known to have raised a question about the pilot's experience was Beverly. She believed that is why Jon Dale emphasized to her that John Fishel nearly had nearly completed his commercial pilot training, an overstatement perhaps offered to appease her concerns. Nevertheless, even if Jon Dale knew the details of the pilot's training, Beverly was convinced that he was determined to go; there would have been no stopping him.

When Sara Brittain finished her talk with Nick Beers, her fears began to run away with her. Off and on since the day the plane went missing, Sara had wondered about the pilot. What troubled her was that she didn't know the pilot personally. Without that personal connection she felt she couldn't gauge his motivation. Remember, several psychics had mentioned that there was something secret or unusual about the flight. Could it be possible that the men had simply left the country, as the insurance people were speculating? Years later, Sara regretted having such un-Christian thoughts, but the thoughts could have easily developed during an emotionally-low period.

This actually wasn't the first time she had a thought that the men had simply flown away. Early in the search, she joined Beverly in a loop drive from Albuquerque through northern New Mexico to put out posters announcing the reward. Bravely or naively, they walked into a darkened bar and asked the bartender whether they could put up a poster. Some of the guys at the bar began to speculate what had happened. One said something like "Oh, these guys got tired of their lives here and decided to take off to Mexico." Sara, full of doubt, did not need to hear this. It started her thinking again that Ken could still be alive somewhere. Beverly reassured her that Ken had no reason to leave her and the kids. He would never do it. Secretly, though, Beverly was hoping in a way that she was right.

To make matters worse, Sara had recently heard that John Fishel and his roommate had talked about going to Mexico on a flying trip. So, Sara phoned Ed Whaley to tell him about her Nick Beers conversation and the Mexico theory. Whaley reassured her that he did not think it was unusual for a pilot to plan flying vacations. Whaley then attempted to change the subject: He passed on the information about the man's cousin who was an 'ESP type' and that the cousin would like pictures of the most aggressive person on the plane, he said. Sara blandly acknowledged the ESP contact information. But something still bothered her.

She came right back to the Mexico possibility. Finally, after months of keeping the question to herself, Sara asked Ed Whaley if he thought the plane could have gone to Mexico? Whaley's notes of the phone call show an emphatic answer: NO. He did mention, however, that other people had asked the same question. Sara seemed relieved, and replied, "I'd bet my life it was not Mexico."

26

Few expected it would come to this. Although some 2,000 hours were flown in the search of the Piper Cherokee and a reward for information on the whereabouts of the plane or passengers was posted, no trace of the plane had been found. An Air Force U-2 photo reconnaissance plane had made flights over the area, big-game hunters combed the area last fall, and hundreds of ground searchers looked, but no positive leads were developed.

Most of the family members had accepted that they would never again see the men alive. However, the uncertainty of not knowing what happened was crippling, as families with soldiers missing in action or with abducted children will attest. Still, the families' drive to continue the search was there and they expected to find answers.

On the one-year anniversary of the men's disappearance, the families gathered together with friends at a memorial service in Albuquerque. A full house of family plus more than 100 friends attended a memorial service at the Eastern Hills Baptist Church, with the Rev. C. E. Allbritton officiating. Wives, parents, and several siblings attended: the parents of the four men were together as a group for the first time. A group photograph of the family members taken after the service shows bright smiles. Rather than being a somber event, the service was a celebration of the men and their lives. Favorite Biblical passages and poetry of the men were read, interspersed with the classic hymns "The Old Rugged Cross," and "How Great Thou Art."

The Rev. C. E. Allbritton spoke of each man and related the love of the outdoors that the four shared. As an example, Jon Dale Horton's favorite poem, written by an unknown cowboy poet, showed his deep pleasure in living in the West. The first part of the poem:

Where the West Winds Blow

"Have you ever lived on the desert plains,
Where the west winds blow and it rarely rains?
Have you ever felt the air grow calm and tense,
As it seemed to hang in still suspense; ..."

Then the pastor turned to three themes that were important for the families to think about. First, don't hesitate to turn to God in a time of need. Second, in good times as well as in times of need, Life has a moral challenge in it. Finally, don't procrastinate about important matters and decisions.

"Tomorrow I will live
The fool does say.
Today itself's too late
The wise lived yesterday."

—Eamonn Fitzgerald's "Rainy Day," Ken Brittain's favorite poem

The Memorial Service served as a nudge to the wives that they should move ahead with their lives. Such was the message of Rev. Allbritton's gentle comment that it is important to not wait until tomorrow to live.

The wives allowed themselves to think ahead. What would be the next steps in their lives? Beverly wrote to her good friends Pat and Joni Henrie in late September:

"I haven't decided exactly what I'm going to do yet. This year has gone and I was hardly aware of the time. Once the baby arrived, I had little choice except to do his will. Now he isn't quite so demanding and I can at least do something without having to make a day's arrangement (for care). I plan to get back in school the first of the year. Where depends on what the requirements are and how much I lose changing States. I would like to finish at Texas University at Austin. That is where Jon went to school. However, if I lose too many hours and the requirements are different, my best bet would be to come back to Albuquerque. But it's a long way from home to be completely alone. So I'll have to weigh it."

Sara also considered returning to school. She was contemplating attending the University of New Mexico and studying music education. Barbara was weighing her options, but as the mother of three young children she was in no hurry to disrupt her newly-established life in Florida.

Whatever next steps the wives took would be made a little easier with news that the Social Security administration approved paying them survivor benefits.

Three weeks after the Memorial Service, the CAP mounted another search for a missing light plane in northern New Mexico. The plane carried four persons and departed the Albuquerque area in late October, 1969 for Denver. While that search was ongoing, the CAP was notified of the discovery of a missing plane in central New Mexico. The latter plane had been missing for nine months and was found by a helicopter rented by the son of one of the men on the plane. The plane had apparently dived into the ground and then bounced into a group of trees.

Just when it would have been easy to give up the search for the Piper Cherokee, the finding of the plane missing for nine months brought hope. *Our plane* and the four men certainly would be found soon.

27

1970

It seemed like an impossible task when the wives filed their insurance claims a little more than a year ago. Many knowledgeable people warned that it could take seven years for all the policies to be rectified. However, soon after the start of 1970, all insurance and social security claims had been paid. Not all claims were paid in full: As predicted by Bill Carpenter, the major insurance companies required the wives to post a bond on the life insurance payments as collateral, in case the men were found alive. The social security claims, in particular, were important because it helped ease the financial burden of raising the children.

The settlement of the insurance claims brought financial security to Barbara Jones. She had moved into her own house in Rockledge, Florida. Her life was consumed with raising her children. Eventually, she worked as a switchboard operator for Brevard Community College. The emotional end of losing Ron was the hard part of her life, but she typically kept the emotions buried deep within.

Beverly Horton returned to art. She wanted to teach, rather than work in commercial art as before. So she enrolled in North Texas University in Denton, part of the Dallas/Fort Worth metroplex, to pursue a Bachelor's Degree in Art History and Education. Once some insurance money arrived, Beverly and her sister bought a house, and her sister's three teen age daughters joined Douglas in the household. Remaining in the Dallas area allowed her to stay close to her family and to the Hortons.

Sara Brittain tried to maintain some normalcy in her and the boys the lives. They were faithful in their church attendance and still lived in the house she and Ken moved into when they moved to Albuquerque in 1967. Inside, though, she was in turmoil. She couldn't understand why God wouldn't permit the plane to be found. She could not seem to give up and turn it all over to God, she said. Her life was at a standstill.

One night after the memorial service she talked with a friend about the missing plane. She shared her doubts, fears, and feelings. Most remarkably, Sara confessed that she believed the plane was somewhere else and Ken was still living. Her friend gently responded, "Sara, I believe the plane's up there and just hasn't been found." For whatever reason, this comment had an impact on Sara. Sara believed that God had used her friend to put peace in her heart.

Thinking that she had turned a corner with a new inner peace, Sara decided she should begin to do something with her life. She enrolled at the University of New Mexico for the second semester of the 1969-70 academic year. Sara already had four years of college and was a good student. Once she started classes at UNM, however, she did poorly in the work, and her grades were not good. She found the studies difficult to concentrate on. She managed to complete the semester, but when she sat down to formulate the coming fall term, she thought "What am I doing here?" She did not return for more schooling. It seemed that the inner peace was short lived; she struggled without knowing what had happened to Ken.

In June, the greening of the high country brought the families back once again for more searching. The focal point was the southern edge of the Pecos Wilderness. Mr. Horton and his party searched around Jacks Creek, while Sara and her group searched the Iron Gate area. Joining Sara was a special guest, Jack Rex. Sara and Jack had recently started dating after being setup by Mary Friberg, and Jack embraced the search with gusto.

Ed Whaley had not heard much from the wives since the memorial service. He assumed they were busy adjusting to their new lives. When Sara Brittain phoned on July 20, she surprised him with the news that she had been put in contact with Peter Hurkos, another prominent psychic. She had an appointment set up with him for July 29 in Los Angeles.

She stated that Mr. Hurkos charged $3500 for his services (equivalent to about $20,000 in 2012 dollars), with no assurance for finding the plane. Whaley was stunned. He advised her not to pay the money, looking at the issue out of concern for Sara's finances. He said that there was nothing for her to gain because all insurance and social security claims had been settled. He felt the money could be better used in a savings account or some investment. From Sara's perspective, however, it wasn't about a gain or loss of investment money. It was simply worth the price to bring closure. Peter Hurkos was a tall Dutchman with a large reputation. While not as famous in America as Jeane Dixon, he was well known internationally for his ability to visually re-create past events by touching articles related to the location. His fame spread as he began to apply these reputed powers to visualizing crime scenes for police departments worldwide. His stature in America grew to the point that in 1960 the TV series "Alcoa Presents: One Step Beyond" presented two 30-minute segments depicting his life and his psychic powers:

> "Peter Hurkos, a Resistance fighter in Denmark in World War II, awakens from a coma and discovers he not only has psychic powers, but that he can see good or evil in someone simply by touching them. Soon he is using his power to track down a crazed killer."

Another indication of Hurkos' celebrity was seen when he was brought in to assist police working the Boston Strangler serial murder case. A wealthy friend of the Massachusetts Attorney General paid to have Hurkos travel to Boston and be thrust into the case. (He did not identify the man eventually charged with the murders). His accuracy has been challenged by critics, who contend he benefits from the 'Jeane Dixon Effect'.

Beverly Horton had read a lot about Peter Hurkos long before the Boston Strangler case. In fact, she had researched his work during the initial search period in 1968 but did not find any way to contact him. Had she been able to reach him two years earlier, he might have been involved with the search in 1968, along with Jeane Dixon and Dr. Gilbert Holloway. Now, she didn't see the need for his help. When Sara called and said she was going to make an appointment with Hurkos, Beverly, like Ed Whaley, tried to talk her out of it. She couldn't. So she told Sara she would meet her in the Los Angeles Airport—she wasn't going to let Sara go on her own.

The women checked in to Universal Studio City hotel in Los Angeles, where Hurkos had reserved rooms for them. Each day, the wives were picked up and had lunch with Hurkos' wife and agent. They then drove to his home for a session, and finally back to the hotel. One of those evenings they got together with John Fishel's aunt and uncle, who lived in Los Angeles and whom they had met during the memorial service. Although Beverly attended the sessions, she stayed in the background and let Sara do the talking with the psychic. The focus of the psychic's work was Ken Brittain.

Notes from Wednesday, July 29, 1970. He started his reading while touching a map brought to him by Sara:

BRILLIANT MIND...DRESSES NICELY BUT AT HOME HE DRESSES SLOPPY AT TIMES. NEVER CRITICIZES PEOPLE MUCH...VERY SINCERE....WON'T SAY ANYTHING ABOUT ANYBODY IF TOLD NOT TO.

WHO'S MIKE - TOM?....PEOPLE WHO WORK WITH GOVERNMENT.

WEATHER BECAME BAD...LEAKING OIL PUMP...TRIED TO CONTACT RADIO.. NOT ENOUGH TIME...WENT TOO FAST.....SOMEBODY HEARD PLANE COMING DOWN...NO RADIO CONTACT...ENGINE FROZE UP...LEAKING OIL...CAME DOWN FAST.

PLANE IN TWO PARTS...PLANE DIDN'T CRASH BECAUSE OF STORM...WAS TURBULENCE AND LEAKING OIL. PLANE SEATS 4...3 MONTHS BEFORE PLANE WAS CHECKED AND WAS IN GOOD SHAPE...NO SABOTAGE

RADIO TROUBLE ONE MONTH BEFORE...ELECTRICAL SYSTEM...WAS NOT A NEW PLANE...WAS AN OVERHAULED ENGINE...WAS NOT A DELUXE PLANE... PLANE BELONGED TO SOMEBODY ELSE BEFORE WAS RENTED.

PLANE BURNED HALF WAY...NO TIME TO RADIO...TOO FAST..WENT DOWN TOO FAST...BRUSH AREA...PLANE HITS MOUNTAINS...OLD MIN...CAVE... PLANE NEAR OLD MINE...FEW MILES FROM MINE...THERE IS A CREEK.

FOGGY AND TURBULENT...ENGINE QUIT...LOSING SPEED...HIT PEAK AND CAME DOWN...THERE ARE NOT 4 MEN TOGETHER...TWO ARE TOGETHER AND TWO ARE OUTSIDE THE PLANE...A MARRIED MAN IN BACK OF PLANE.

PLANE IN TWO PARTS...YOU STILL CAN SEE PART OF PLANE...3 OR MAYBE 4 MILES FROM OLD MINE (CLOSED) SHACK NEARBY...SOMEONE HEARD PLANE IN TROUBLE...ROAD NEAR MOUNTAIN AND THEY HEARD IT...NO SABOTAGE...ALL DEAD...NO FLESH ON BODY...BONES ONLY...

WHO HAD WINDBREAKER TYPE JACKET ON?....WHAT DID THEY HAVE TO DO WITH ELECTRONICS?...I SEE AN ELECTRONIC TECHNICIAN...PEOPLE ARE LOOKING IN WRONG AREA FOR PLANE.

THEY TURNED BACK...TROUBLE STARTED HERE ....(Peter points to area on map.)

BUSHES...NEED HELICOPTER AND FOOT...CAN'T FIND PLANE BY REGULAR AIRPLANE...YOU NEED HELICOPTER.

ALL ARE KILLED INSTANTLY...NO PAIN...PLANE CAME DOWN AT ABOUT 140 M.P.H. SOMEONE HAD A SMALL CASE... LIKE FOR CARRYING MAPS...DARK COLOR.

Sara and Beverly left the session half-exhausted, half-intrigued. Later that afternoon, Sara phoned Ed Whaley and gave him a synopsis of what was said:

Plane search has been conducted in the wrong place (lower Pecos country).
Plane in two pieces. It had oil pump trouble and went down at about 140 mph. Turbulence contributed to crash. Plane was partially burned when it struck the ground.

Sara asked Whaley to air-express a set of detailed topographic maps. They reconvened the reading the next morning.

Thursday, July 30, 1970

"WHEN ENGINE TROUBLE STARTED THEY TRIED TO PICK UP SPEED...PILOT NOT GOOD FOR MOUNTAIN FLYING...GOOD FOR FLAT LAND BUT NOT GOOD FOR MOUNTAINS.
AT 10,000 FEET THEY WERE IN CLOUDS FOR 7 TO 9 MINUTES. THEY TRIED TO FIND LANDING STRIP...4 OR 5 MILES AWAY THERE WAS A STRIP TO LAND... PILOT KNEW NOTHING ABOUT NAVIGATION. PILOT DID NOT FLY EVERY DAY... HAD ANOTHER JOB AND THIS WAS JUST A HOBBY FOR HIM. NOT EXPERIENCED ENOUGH.
PILOT THOUGHT HE WAS ABOUT 11,000 FEET (IN CLOUDS) EVERYTHING SMOOTH THEN TURBULENCE...ENGINE LOST POWER...OIL DEFECT...LOSE ALTITUDE. TRIED TO PICK UP SPEED...DIDN'T HAVE A CHANCE TO GET DOWN TO VALLEY...HE TOOK THE WRONG SIDE...THOUGHT IT WAS NORTH AND IT WAS WEST.
PLANE IN TWO PARTS...TAIL IN SHAPE...FRONT BURNED UP...FRAME ONLY...8,900 FEET. IT HIT MOUNTAIN...6 TO 9 MINUTES IN CLOUDS.
MR. BRITTAIN ALWAYS QUIET...HOLDS IN STUFF...WHAT'S WRONG WITH MOTHER'S LEG? ...TROUBLES WITH LEG...I SEE BLUE SPOTS...CALM FAMILY.
TWO IN PLANE...2 OUT. ONES IN BACK ARE OUT...4 MILES FROM AIR STRIP....A SMALL STRIP...DIRT.
PLANE TOOK OFF AT 7 AM AND NOT 6:40 AM

The session was a mixed bag—some new findings but many findings that came forth the day before. The maps sent by Whaley arrived in the morning and the wives turned them over to Hurkos for study in the evening.

In terms of guiding them to where the plane was, nothing solid had yet developed. With regard to Peter Hurkos' psychic abilities, it was difficult to say. The flight patterns he spoke of could never be verified. The statements about Ken Brittain's personality and habits likely could have been gleaned from careful study of Sara's behavior. The finding that there would only be bones left at the crash site is common sense and what one would expect after so much time. Confirming the cause of the plane crash and the condition of the wreckage would require an actual inspection of the aircraft, if it was ever found.

There were, however, a couple of intriguing points in Peter Hurkos' readings. The first pertained to his observation that Ken Brittain's mother suffered from troubles in the leg—blue spots. In fact, Mrs. Brittain in November 1968 developed gangrene in a toe due to sugar diabetes and there was a possibility the toe might have to be amputated, though it was not. He also provided Beverly information about her mother's illness that was spot on accurate. The second point came during the day when he had Sara give him a photograph of one of her children. The photo was given to him face down and he described the photo by looking at the back. Such a psychic ability was also suggested in the *Boston Strangler* movie. Lastly, there was the somewhat strange event when Hurkos phoned the wives at midnight and asked them, "Who is Jones? Jones was sitting in the back." Of course, Ron Jones was a passenger on the plane. Supposedly, neither Beverly nor Sara had told him about Jones. Whether this was indicative of psychic ability or not was debatable. The identification of a person named 'Jones' could be accounted for if Hurkos had seen any of the extensive news coverage about the missing plane.

His mention of those health issues and of Ron Jones caused the wives to wonder, but they were not ready to extol his psychic abilities. They had been very careful about giving him information during the lunches or sessions; they had come for information they did not have and agreed not to furnish him with anymore than they had to. The wives did, however, talk amongst themselves in the hotel room. They wondered that since their rooms were arranged for them, might the psychic's operation be gathering information on their lives?

Friday's session with Peter Hurkos was shorter than the others. He again stressed that it is very necessary to use a helicopter for the search.

Then Hurkos suggested a general location where the aircraft was—west of Truchas peak and west of Tres Ritos. Ed Whaley's reaction to this was that the location does not seem to be where the plane should have been flying.

Sara and Beverly returned for one final meeting with Peter Hurkos on Saturday, August 1. The tall Dutchman laid out a topographic map and proceeded to provide the most important piece of information to the wives—the location of the plane. He said the missing airplane was in an isolated river valley near the northwest corner of the Truchas Peaks. Specifically, he pointed to a segment of stream below the San Leonardo Lakes. With a felt-tipped pen, he highlighted a narrow, two-mile long segment adjacent to the stream, between the elevations of 9,000 and 10,000 feet. He added arrows that indicated the direction the plane was flying before crashing. According to his sketch, the plane was flying up-canyon directly toward the peaks. Finally, Hurkos drew a small box showing the precise location to find the wreckage.

Sara and Beverly flew together back to Albuquerque. Two days later, Beverly Horton checked in with Ed Whaley. She said that she and Sara tried to rent a helicopter to search the Rio de San Leonardo, but they found none would be available for two weeks. Beverly noted that Peter Hurkos had stated the plane would be found in two weeks. Because of her family commitments, though, Beverly needed to return to Texas and couldn't wait around until then.

Shortly thereafter, a group of men from the Search Team hiked into the San Leonardo Lakes area and headed directly to the location marked by Peter Hurkos. Ralph Grill, Jim Kohl, and Ed Whaley thoroughly combed the stream bottom and nearby hillsides. No evidence of the plane was found. By coincidence, the same drainage was flown a day earlier by a State Game and Fish helicopter looking for a lost Boy Scout, and they also did not spot a plane.

Three weeks after saying goodbye to Peter Hurkos, Beverly hopped aboard a rented helicopter and flew to the San Leonardo Lakes area. The helicopter stayed up a couple of hours before bad winds forced it to return. The helicopter searched the same approximate area that the ground searchers and the Fish and Game helicopter looked at before. Sara, Jack Rex and his family and friends drove to the site and hiked the area.

A couple days later, Beverly joined Sara, Jack and his family on another ground search of the Hurkos target area. Adjacent to a beautiful grassy meadow within the valley floor was a dense forest, full of huge fallen trees that the group had to crawl over or under. It was almost impossible to maneuver, with little light, and Beverly once had to use a compass to find her way out of the woods.

No plane was found. The negative results continued to indicate that the psychic reading was false, neutralizing the electricity that the wives felt after being told exactly where to look.

Sara spoke by phone with Peter Hurkos on two more occasions. He still maintained the plane was in the San Leonardo Lakes area.

28

Leaving no stone unturned in her quest to try anything for a successful search, Sara Brittain turned to the press in early September, 1971. The headline in Sunday's *Albuquerque Journal* read, "Cargo's Aid May be Sought in Search". Her interview with the reporter said that she may appeal to Governor David F. Cargo for help, through the National Guard, before winter snows once again cover the rugged Pecos Wilderness.

"We've camped, walked, hiked, backpacked, we've done everything possible. We've tried to get off the trails where nobody would go. We've done it all. We spent about two or three hours in a helicopter up there three weeks ago. Even the people in the helicopter had trouble seeing the ground in that dense country. It would be real hard to spot anything up there. Representatives of all the families came out here last July and spent about two weeks in the Pecos area, and then friends spent I don't know how many other hours looking. It could be anywhere up there. We don't have any clues and we just don't know. If we could say for sure it's in the Pecos, then it would be okay. But we just don't know for sure".

The article recalled that former Govenor Jack Campbell faced a similar situation in 1965 when three other Albuquerque men were missing. Campbell approved the use of Army National Guard reconnaissance planes, and their missing aircraft was found shortly thereafter. Dave Cargo could not yet respond to such a request because he was on an official visit to Ireland. The article's pre-emptive timing and the mention of Campbell's success, though, left Cargo little room but to respond publically. The reporter said that the request for help may be made by "The shy and attractive woman as a last resort."

Within a week, Governor Cargo returned from his overseas trip and met briefly with Sara. He did say that he would ask the U.S. Air Force to take special aerial photographs of the Pecos Wilderness. Then, with characteristic bluntness, he told the *Albuquerque Journal* that "I could sympathize with her, but all we would find would be a wreckage and no one alive. I did explain to her that such an operation would be expensive—up to $50,000 to $75,000 per day."

The two-year anniversary of the October 2 plane disappearance passed with little fanfare. Two weeks later Sara decided to check in with Ed Whaley. The chat was casual as neither party had much news to share. She reported that, to her knowledge, no additional aerial photos were taken or planned; state agencies were again asked to be alert.

There was only one real item of news: Jack and Sara were going to get married. Jack had proposed twice before she agreed.

Sara and Jack were hoping for a Thanksgiving wedding. But before they could get married she would have to petition to have Ken Brittain officially declared dead by the courts. The ruling came down on November 23, 1970 from the offices of the Honorable Gerald D. Fowlie, New Mexico District Judge:

> "...It is, therefore, ordered, adjudged and decreed that Kenneth Paul Brittain died on October 2, 1968, within the State of New Mexico and that his death was proximately caused by the crash of a private aircraft and that his death be, and it herby is, judicially established.
> 'It is further ordered, adjudged and decreed that that the Department of Health and Social Services, Division of Vital Statistics, State of New Mexico, be, and it hereby is, ordered to issue a certificate of death of the said Kenneth Paul Brittain in accordance with the provisions of this Judgment."

The Thanksgiving wedding of Jack Rex and Sara Crouse Brittain was a low-keyed affair, attended by immediate family and close friends.

The last remaining 1970 entry in the Gulton Industries files was made by Morys Hines on October 21. Hines was called by Sara who stated she had talked to an unnamed woman clairvoyant in Albuquerque who "pin-pointed the location of the downed plane." The location was in the general area that

Dr. Gilbert Holloway stated, around Wheeler Peak. The clairvoyant said that the pilot knew of a landing strip in the area and tried to land but could not see due to mist (fog). Sara noted that Peter Hurkos earlier had mentioned a landing strip around *his* area. Sara's checking of the maps showed a landing strip near Wheeler. No one from Gulton was available then to investigate the clairvoyant's readings, and it is not known if the families pursued this further.

It is likely that the folks at Gulton Industries suffered from "search fatigue" at this point. After aggressively pursuing the Peter Hurkos predictions, the company appeared to have accepted the possibility that the plane would only be found by accident. Two solid years of searching had been dedicated to the effort. One can only hope to work for a company that would do the same.

29

1971

Ed Whaley continued to be amazed at how many different people offered to help, or at least offered advice, in the search for the four missing men from Albuquerque. Whenever it appeared that the public had forgotten about the plane, along came someone—seemingly out of the blue—who made the effort to connect with the Search Team. Sometimes the ideas or suggestions were borderline nutty, but Whaley always appreciated the intent.

The search season of 1971 started off for Ed Whaley in mid-June with a familiar theme: ESP types (as Whaley referred to them). After all the disappointments so far, Whaley continued to treat each prognostication with as neutral of response that he could muster. The latest offering came from a local welder who just returned from serving in Vietnam. While on active duty, the welder spoke to another soldier who had a dream about this particular plane crash. Because the person described the plane in accurate detail, the welder took great interest in the missing Piper Cherokee. He visited with Ed Whaley and borrowed several binders of search notes of the search to study. Nothing further came from the dream.

A week later, Whaley got a phone call from Mr. John Horton. Whaley was happy to hear from the man and was touched by his determination. Once again, the families would return to New Mexico to carry on the search. Mr. Horton was to meet up with the other fathers, Mr. Jones from Florida and Mr. Fishel from Nebraska. Mr. Brittain occasionally was in town on business and might also join them. They would spend two weeks searching along the northernmost leg of the flight plan. As with previous efforts, the men would encounter thickly wooded mountains.

Mr. Horton shared with Whaley some troubling information about the sightings in the Pecos area. Last Christmas he received a letter from Mrs. Smith, the camp caretaker who heard a plane the morning of October 2, 1968. She now believed the plane she heard was not *our* plane, but a plane belonging to a friend of hers. It seems as if she had been talking to a friend who flew up the Pecos Canyon that same morning. Immediately, Whaley wondered about the sighting of Stanley Gonzales. Was this the plane Stanley saw? With this revelation, Whaley sighed. Back to ground zero!

A caravan of Texans arrived early June. The Hortons came with their camper and jeep, followed by a family friend and his teenage son, Eddie. Beverly drove her car. After picking up Mr. Jones who flew in from Florida, the group headed to the high country. They moved Pop Tucker and his animals to a park that had a corral and water, just above the village of Cowles at the end of the paved road from Pecos.

Spirits were high, as the third year of searching got underway. However, those sprits were dimmed in just a few days. Pop Tucker got hurt by a mule kick and couldn't take care of his animals. The others tried to help with the animals, but couldn't make up for him. Without Pop Tucker, the group's ability to get around was severely limited. No one trusted the mule anymore, and they didn't want Beverly and Eddie to be out there alone on the Jennies. To make matters worse, it rained for almost two weeks. By the end of the two weeks, they moved Pop Tucker and his animals back to Albuquerque. Beverly and Mr. Jones returned to their homes. The Hortons moved to a more comfortable location and stayed around the Wilderness for another month.

On July 1, Whaley's normally busy work day was pleasantly interrupted by a visit by Mr. and Mrs. Jack Rex. He was happy for Sara's new marriage and this was really the first time he had a chance to sit with her husband Jack. Jack announced that they were still interested in the plane search. He recalled for Whaley the efforts the family had made searching the Peter Hurkos site. The couple now did not believe the plane was at the site. It was revealing when Sara said she did not expect much from the ESP types. "I have learned my lesson," she concluded. Their chat ended with

Jack mentioning they were interested in a report made last winter about a hunter who saw a downed plane near the Colorado border. Whaley recalled the incident and said that the Albuquerque CAP had searched this area extensively. Nonetheless, the family would shift their efforts to looking for the Piper Cherokee in this area.

As the meeting drew to a close, Ed Whaley said he had some news of his own to share. He announced that he was transferring to Pennsylvania to serve as Executive Vice President of a different division of Gulton Industries. He would be leaving behind the Data Systems Division, which he dearly loved and had nurtured since 1958 into a national technological powerhouse. In fact, much of the software and hardware in the mission control facilities at NASA's John F. Kennedy Space Center were provided by Gulton's Data System Division. The news was the last thing Sara expected. The search truly was a true family affair now, with much of the burden falling to Sara and Jack.

It would have been understandable if Jack wasn't enthusiastic about spending his weekends searching for the body of his wife's first husband. However, Jack embraced the task as if one of his own family members went missing. Having witnessed Sara cry herself to sleep more than once over this, it was clear that this was something he had to do.

Jack Rex was a bundle of eclectic energy. He loved to talk. He loved to discuss, and usually complain about, politics. He loved to run. He loved to explore the woods and photograph ancient rock inscriptions. He became fascinated with the different forms of barbed wire and over time amassed an extensive collection. And on the holy day of Sunday, he would follow worship at church with his sacred duty to watch his beloved Dallas Cowboys.

For now, Jack channeled that energy to the search. Coming into the effort several years later than others, he brought a naivety that this would be straightforward: Spend a few months driving back roads and the plane should be easy to spot.

He started his education by talking with Earl Livingston of the Civil Air Patrol. After the preliminaries, Livingston told a story that a hunter last winter saw a blue and white plane near the tiny town of Amalia, New Mexico, just a few miles from the Colorado border. Supposedly, the plane was lying on its side with no wings. The hunter saw a "180" with a circle around it, which would probably identify it as a Piper Cherokee. The problem was that the hunter saw the plane as he ran by it, following a deer he had wounded, and he did not bother to retrace his steps to examine the wreckage or mark its location. In addition, the CAP could never identify who the hunter was or verify the second-hand story. They did aerial searches of the area but didn't spot any wreckage.

So Jack had his first mission—track down the hunter. It took him two weeks of phone calls and interviews to identify the hunter. The hunter told Jack that he and a group of hunters went up to Amalia early in the morning and split up into smaller groups for the hunt. Alone on a ridge, he looked up and saw a blue and white airplane, with no wings, and the nose was black, possibly burnt.

Jack then spoke with the others in his hunting party and they remembered passing two shacks, an outhouse, and corrals before heading up a canyon in a jeep. They went to the top of a ridge at the end of the canyon and hunted from there. After the hunter told them of the plane, the party returned the next day and they were unable to find the plane. It would be one of Jack's prime spots to search next year.

30

1972

Beverly Horton had no desire to re-marry and so entering the work force again was critical for her and Douglas' future. She finished her studies the spring term of 1972, and immediately began to look for a job. She accepted a teaching job in Holbrook, Nebraska as a high school Art teacher, and spent the rest of the summer getting ready to make the move. Even though she left her immediate family, her desire was strong to be independent and to provide for her son's needs.

During the fall, Beverly started studying part-time for her Master's Degree in Education at Kearney State University.

The winter months crept by for Jack and Sara. In this time, Jack took on the chore of compiling an inventory of all known airplane wreckages that were scattered throughout the state. According to Jack, the Civil Air Patrol did not maintain such as list, partly because they were not involved in responding to military crashes. There were numerous times during this search when wreckage was spotted and field teams deployed, only to discover it was from an older crash. Although the Air Force Rescue Center supposedly held such a list, Jack felt it was not very accessible and he wanted his own to share with the CAP. Besides, Jack was not one to idly sit around until the field season started again and this gave him something to focus on.

When not working on the list, Jack also closely followed newspaper stories of other missing airplanes in the West and any passenger survival stories. There was plenty to keep his attention: In the span of two spring weeks, 12 persons died in three separate plane crashes in New Mexico.

On the last day of March, 1972, a letter was printed in the *Albuquerque Journal* in which another 'sighting' was described. The writer said that her son-in law and his party of hunters saw the wreckage of a plane near Amalia, New Mexico, close to the Colorado state line. The party couldn't get up to the wreckage because of the deep snow, so they could not say what kind it was. The writer was asking the newspaper for help in locating the "women who lost their husbands in the plane wreck almost four years ago." The newspaper forwarded the request to Earl Livingston of the Civil Air Patrol, who in turn contacted Sara and Jack. This was the third recent report of airplane wreckages being spotted near Amalia. Jack directly contacted a member of the hunting party, who was not so sure what they had seen: they saw something shiny in a ravine, but didn't want to get anybody's hopes up. Toward the end of April, the hunters slogged through five feet of snow to return to the site and found pieces of an old DC-3 plane crash. Wrong plane!

The Albuquerque newspaper byline the first weekend in June announced that the "Plane Hunt Resumes Monday". The parents of three of the missing men arrived in town, like clockwork. John L. Horton, R. E. Jones, and Mr. and Mrs. Arthur Brittain spent two weeks of intensive searching the canyons and ravines of northern New Mexico: a quixotic attempt in the vast high country.

As the fourth anniversary of the plane's disappearance approached in early October, Beverly realized that hunting season was approaching and she wanted to raise the Reward Fund. She had a job. She had more money. So Beverly called Sara with a proposition. If she sent Sara a check for $500, could Sara get it added to the Fund and get a story to the newspaper? Sara said that if Beverly sent the money, she would find another $500 to match. So by the fourth anniversary, the Rescue Fund made it to $2000. "Hunting season is coming up and we feel now is the time to remind people that the plane is still missing," Jack told the papers.

A few weeks later, Sara sat with an *Albuquerque Tribune* newspaper reporter for another front-page story about the missing plane. Although four years is an eternity in the news business, this story was still very compelling. As Sara sat in her kitchen, she recalled the phone call from Beverly Horton the morning of October 2, 1968.

"She told me they hadn't heard from the plane—that it was overdue. That phone call changed my life."

For about a year and a half after the plane was reported missing, Sara couldn't believe her husband was dead. She hoped and prayed.

"I refused to give up. I thought they would be okay. There is something in me that refuses to give up."

"But that seems like another life. I had to quit dwelling on the past. Then I decided I had to look toward the future. If it could be found, we could breathe a sigh of relief."

Sara and Jack continued to pursue leads—something in the trees, reports of someone seeing a low-flying plane in trouble, a burned place in the forest. So far, nothing. But, one never knew when *the* lead will come in. A recent news clipping from Colorado reinforced this to the couple: "Plane Wreckage, Body Found After 13 Years."

While the mountains were quieting down before the first major snows, search activities were picking up around Sara and Jack's household. This latest news cycle brought letters and calls from around the country. The first lead came from a house painter from Española. The painter said he was with a group elk hunting north of Chama in December, 1968—nearly four years ago—when he saw the tail of a plane between two trees. "The plane was white and light blue," he told a reporter. There was snow on the ground and only about four feet of the tail was visible. Because it was snowing hard, he left the site without looking further and returned to camp. The group later shot its elk "so we had no reason to go back to the plane." The painter said he reported the incident to the State Police but they weren't interested. "The policeman said the plane they were looking for disappeared over the Pecos Wilderness." He contacted Jack after the recent news accounts, in case this might be helpful.

The second lead came from a Taos potter. The potter spotted a light-colored plane in the Pecos Wilderness one year ago while on a backpacking trip with his family west of Truchas Peak. "It was in a wooded gully or a ravine," he said. Although the family got within 25 yards of the wreckage, they went no further because "we had no interest in the plane whatsoever, so we went on." When the potter read Jack's plea for information in the *Tribune* newspaper article, he contacted them. He guessed it would take two days on horseback or foot to reach the wreckage.

A Phoenix upholsterer provided the third new lead. He was elk hunting with friends in October when he spotted two streaks of white metal about 12 feet long. Later, a friend told him about the missing plane and he contacted Jack Rex. He said the plane might be in water, but he was one-half mile away when he spotted the plane and didn't get closer. Mr. and Mrs. Brittain, Ken's parents in Phoenix, interviewed the upholsterer and found him to be "sincere in thinking he saw the objects." The upholsterer had prepared a map for the Brittains based on a hand sketch he made with a bullet during the hunt. The location was near the Winsor Trail, perhaps close to Spirit Lake or Lake Katherine. It would take a couple days of hiking to get to the place.

The most unusual tip came from a dowser from Pennsylvania. He wrote to Jack:

"We have already found that by going over a map or sketch of an area with a pointer, as if walking there, a spot will be indicated by our rod where an object can be located. Imagine how many lives might be saved in a crash with this quick method if it would only work every time."

The dowser wrote that he had located a plane crash in Washington State and a rich deposit of copper in Colorado with this method.

"So you can see why I keep trying to find things this way, always hoping to be of help to someone. There must be some unknown natural laws governing this strange gift which we as a society are trying to discover."

Included in the letter was a hand-drawn map with a mark showing the likely location of the missing plane. It was placed about twenty miles north of Truchas Peak.

"Please let me know if this spot I've found is anywhere near the plane's finally discovered location. So many people just ignore or ridicule this sort of thing they can't understand."

They were a smorgasbord of leads. Sara was overloaded with possible places to search: Chama, north of Truchas Peak, west of Truchas Peak, Winsor Trail. She and Jack would consider which lead to pursue over the winter months.

The couple was looking forward to a rest from search activities. However, this interlude was short-lived. The November 14, 1972 edition of the *Albuquerque Tribune* drew gasps from Sara and Jack when they opened the afternoon paper:

Plane lost 4 years believed found at Golden

The wreckage of a light plane fitting the description of one which disappeared four years ago with four elk hunters aboard was spotted today near Golden. Air and ground searchers were moving into the hilly and brushy area about three miles south of Golden early this afternoon to identify the wreckage. Early reports said that human remains were found in the wreckage. Golden is an old Santa Fe County mining community about 35 miles northeast of Albuquerque. State Police Capt. Ernest Tafoya said a hunter, identified as Ernest Thompson, reported finding the wreckage between three and four miles south of Golden. The area is sparsely populated.... The site where today's wreckage was spotted was about 50 miles from the Pecos Wilderness, indicating the crash occurred either as the men (elk hunters) were starting out or on their return....

The optimistic headline provided Sara with a surge of hope that maybe the nightmare was over. There still remained a small possibility, though, that the wreckage was another plane.

Unfortunately, any relief that Sara felt was dashed by the end of the day. It took little time for the recovery team to discover that this wreckage was that of a different crash, one that was found earlier in 1969. "The number we got off the aircraft was not near the number of the plane that had been missing so long," an Air Force official said. A CAP member said he recognized the plane and location as being another. The body of the man killed in the other wreckage was removed in 1969. There were additional bones found at the site by the current 1972 rescue operation, but it was unclear if they were human.

The headline in the *Albuquerque Tribune* raised hopes with the families. It appears the paper had the best intentions and wrote the headline because it was excited for the families and what seemed to be a resolution of the four-year mystery. The next day, stories in both local newspapers expressed regret that the missing aircraft still had not been found.

PART IV

Discovery

31

1973

It was springtime. In most years, Sara would have already touched base with Ed Whaley and Beverly to plan the search for the coming summer months. Certainly there was no shortage of possible search locations. Just last fall, at least four new clues came in to pursue. In the past, Sara and Jack normally would follow up on these leads in some way, and the families would spend time in the mountains looking for the wreckage. Every lead was vigorously pursued as best as the families could, with their limited numbers and resources.

For the past four years, Sara felt that she couldn't relax or accept the uncertainty. Finally, in 1973, she began to feel uncomfortable about the situation. She was tired of trying. She decided not to pursue those leads. There would be no more searching for Jack or Sara. It was a monumental shift in direction.

After making their huge decision not to search further, they decided instead to plan a summer vacation—Sara's first in nearly five years. Rather than hiking the high country looking for the missing plane, the couple would be on vacation.

During the years since Ken's disappearance, Sara had not been together with all of her immediate family at one time. Of course, her family had visited New Mexico periodically during that time, but they always came in small groups. Sara missed the large gatherings. Consequently when Sara and Jack began to plan their vacation, they first thought of traveling back to Illinois for a mini-family reunion. However, when they talked with her family, plans changed. The consensus from her family was that they'd like to come to New Mexico and do a road/camping trip together.

ALBUQUERQUE, NEW MEXICO, THURSDAY, AUGUST 2, 1973

As she walked into her bedroom, Sara hesitated for a moment after she opened the closet doors and spotted the books that were bound together by rubber bands. She clutched the diaries that had captured her life's thoughts since she was a child. Finding the most recent book, she opened it to the last entry and realized that it had been since Christmas, 1968—nearly five years ago—when she last felt like writing anything in her diary.

Things were different now. Her mind was ready to open up again. The changes that had come over her recently would hardly be perceptible on a day-by-day basis. Without recognizing it, she had turned a corner. Sure, she had remarried. And, her immediate family would soon be united again on vacation. However, it was her decision to "turn the search back over to the Lord", that gave her unexpected happiness and peace. When God wanted the plane to be found, she believed, it would happen.

HONOLULU, HAWAII ,WEDNESDAY, AUGUST 1, 1973

Beverly Horton's exuberance was in high throttle as she and Douglas arrived at the Honolulu airport. Certainly, it helped being in such a beautiful place. It was the warm hug she received from her sister, though, that was most important. She looked forward to playing on the beach with Douglas, shopping for colorful art work, and relaxing with good conversation.

As she watched the sun set on the Pacific Ocean horizon that first day, she sighed. A trip made in Heaven. This vacation represented a transition for Beverly.

While the school year in Nebraska had drawn to a close, Beverly Horton was feeling tired and was no longer motivated to return for the search. Now that she was a full-time teacher, she especially valued her time off. Most importantly, she wanted to relax and have fun with her son. Her sister's phone call gave her an answer as to how to spend the summer months: Would Beverly and Douglas consider joining them for a vacation in Honolulu? Consider? We accept! Her sister had lived in Hawaii ever since Beverly was in college; Beverly's first trip to Hawaii was in 1964 and she had visited many times after that. This would be Douglas' second time visiting the islands. They had planned to stay three weeks.

Apart from just spending time with her sister, the trip had an added benefit of some free child care. Beverly's brother in-law had offered to entertain Douglas while the sisters explored some of the other Hawaiian Islands.

Maui, Hawaii, Friday, August 3, 1973

After a few days in Honolulu, Beverly and her sister decided to fly to Maui and explore. Beverly's sister worked for an advertising agency and one of its customers had given her some perks for a stay at a hotel in Maui. Added to the hotel perks, were cheap airline tickets. In those days, if you had purchased an airline ticket from the mainland to Honolulu, for just $10 each way you could island hop to any you wanted. They decided to take Douglas along on the trip, because Beverly's brother in-law couldn't watch him due to work commitments over the weekend. When the sisters arrived on the island, they rented a small car and made the rounds of the island, whale watching and the like.

That evening, they got a babysitter at the hotel for Douglas and planned on having dinner and touring shops and other activities at the hotel. During dinner, her sister asked the oft-avoided question of how Beverly was coping. For a moment, Beverly almost gave her standard response that she and Douglas were getting along. She paused and then began to tear up. A completely different answer came forth.

"I would like to live my life where it is not last thing I think of before I go to bed, and it's not the first thing I think of when I wake up," she said between sobs.

"I just need to know. I really need to know...." Struggling to finish her answer, she continued, "I'm just going to just have to give it up. I've concluded I'm just going to have to give it up."

Her sister knew that "give it up" meant she would give up the search. Beverly thus became the final wife to stop actively searching for her husband. Perhaps it was due to the drain of the never-ending search or to the relentless uncertainty, but Beverly felt emotionally spent and needed to let things happen as they may, and move on with her life. Like Sara, Beverly had made the decision to allow fate or luck take over. She thus followed Barbara Jones and Sara Brittain in giving up the search.

This was a major shift in Beverly's determination. It was the first time her sister didn't see that steely drive that kept Beverly going. Her sister tried to think of something comforting to say to Beverly but nothing novel came to mind, so she just nodded her head in agreement and in support.

32

Pecos Wilderness, New Mexico, Saturday, August 4, 1973

We decided to go backpacking into the Pecos just a few days before. This was to be a short trip, in on Saturday, out on Sunday. So, we had few major expectations for the weekend. We didn't know exactly where we were going, but we'd figure that out as we went. To us, the destination was not as important as simply being outdoors. We would hike to a beautiful spot and enjoy the wilderness.

I had returned to my hometown of Los Alamos, NM at the start of 1973 to look for work following college, and figure out what my next big step in life might be. Good fortune came to me not too long after my return. Along with landing an interesting job, I became reacquainted with Janie Hones, a fellow high schooler. I ran into her at the gym one evening after work and called her at home the next day.

"Could we get together, if you're available and willing?" I asked.

"I'm available, but not necessarily willing", she teased.

Janie and I hadn't known each other well in high school because she was a late transfer to the school. In 1973, our friendship blossomed.

Janie was relatively new to backpacking but evidently liked it because this was to be our second trip into the wilderness in a month. During Janie's inaugural backpacking trip, we scrambled to the top of Truchas Peak, the second highest peak in New Mexico, with a large group of friends from our High School class. In spite of her compact size—standing maybe five feet and barely topping 100 pounds—I soon discovered that she was a dynamo hiker with piston legs. Besides, she was cute and had an infectious laugh.

On a short two-day trip such as this, you can carry heavier, more delightful food than for an extended venture—no freeze-dried food for us. So, at the grocery store, the essentials of chocolate candy bars, cheese bricks, fancy crackers went into the cart. But when I passed the meats section, my eyes locked onto a huge sirloin steak. Yes, Janie will thank me for carrying this along, I thought.

I drove home and placed the steak in the freezer; the frozen meat then was buried deep within my backpack and hopefully would stay fresh until dinner. A rain poncho, Frisbee, utensils, cooking pan, flashlight, and so on went into the pack. I decided to leave my cook stove at home and grill over a live fire—it would be almost criminal to cook such a steak over a cook stove. For shelter, we used a tent that my friend Lou kindly loaned me. Although it was a risk to borrow equipment, the tent would be fine, I convinced myself: it hadn't rained in weeks and we mostly needed it to keep the bugs away.

We left the car in the parking lot at the base of the Santa Fe Ski Basin, wrestled our packs on, and made our way up the top of the cleared ski runs toward the boundary of the wilderness area which bordered the ski basin. Although we were young at age 23, we took our time and let our bodies adjust to carrying a load up steep terrain in thin air. At over 10,000 feet, the altitude at the base of the Santa Fe Ski Basin is one of the highest in the nation. Along the way, we stopped and took a quick nap among the grasses and wildflowers. Soon we entered a thick spruce-fir forest and in little time found a hiking trail that probably started as an ancient game trail.

In a half hour we emerged from the upper limits of the forested area—timberline—and the trail path was marked with piles of rocks called cairns. The sky was brilliantly clear and the trail beckoned us to continue to the summit of Lake Peak, at 12,049 feet.

Except for the Lake Peak summit, we didn't have a specific destination for this trip. After reaching the summit, we would choose our path as explorers of the area. We would rely on our intuition, along with knowledge of the area, and map and compass to decide if there was a safe and obvious hiking route over the peak. Nearly all visitors to Lake Peak are day hikers that turn around at the summit and return to the Santa Fe Ski Basin.

Normally, if one wanted to get to the interior of the Pecos Wilderness in this vicinity, a hiker would not follow our path to Lake Peak, but would instead follow the Winsor Trail from the ski basin. The Winsor Trail is among the most heavily used trails in the Wilderness, and few people contemplate the alternative of hiking into the wilderness via Lake Peak. If successful in navigating off Lake Peak, we would be among the small number of recreationists that continued their trip past the summit.

Nestled in the shelter of some rocks at the summit, we saw two prominent land forms nearby: Santa Fe Baldy and Penitente Peak. To the north was Santa Fe Baldy, a massive near-circular rampart that is some six hundred feet higher than Lake Peak, and the highest point overlooking Santa Fe. Glaciated, very steep slopes were on the west and north face of Lake Peak so it was not feasible to hike easily toward the direction of Santa Fe Baldy.

The other feature in view was Penitente Peak, a land form attached to the east side of Lake Peak by the thread of a narrow ridgeline. In contrast to the rugged alpine peaks in the vicinity, the shape of Penitente Peak was unique. Penitente is more of an elongated mesa, resembling a long bulbous nose, with open grassy slopes wrapping the ridgeline of the peak for nearly a mile to the north. As we studied the landscape, we noticed what looked to be a faint trail that curled around the nearest edge of Penitente. Where that trail led we did not know, but, in keeping with the explorer theme, we decided to cross to Penitente Peak and follow its path.

However, after reaching the edge of Penitente Peak the trail abruptly ended only 100 steps later. Partly covered by grasses along its edges, it was obvious that the trail was mainly a game or livestock path and rarely used by humans. I was inclined to stop our exploring here at the apparent end of the trail, having never hiked on Penitente Peak before and feeling responsible for Janie's well being.

Because we had no firm destination, I thought it can't be considered to be a failure to stop here: we can always camp nearby for the night and hike back to the car tomorrow via Lake Peak. Beautiful country, great company, and we certainly had the place all to ourselves—it had been more than four hours ago since we had last seen another person. Before deciding what to do next, we removed our packs and sat down near the trail to study the topographic map. With our backs leaned against an open grassy hillside, we had a good view of the trail in both directions.

The man appeared seemingly out of nowhere and without a sound. "Where are you headed to?" he asked.

Somewhat startled, I looked up from the map and at the man. He was standing about 10 feet off, diagonally in front of us. I couldn't see his face well because he was backlit by the afternoon sun. Dressed in plaid wool shirt and wool pants, he had a sandy blonde beard. He appeared to be dressed as a mountaineer might. I chuckled to myself that he must be an angel, appearing out of nowhere and so brilliantly lit up.

"Well, I'm not sure," I answered. "We're debating that now," as I held up the topographic map.

He smiled and said "I've got just the trip for you". He turned and pointed to the north. "Head up that hill (Penitente Peak) and you will find a primitive trail. Don't confuse this with the big trail—this one is faint. It isn't on the maps—they've taken it off the maps to let it grow over. Stay on the trail and you'll get to the top of the mesa." Then he warned us: "Eventually, the trail will go in and out of the trees. At some point you might lose the trail. Don't be worried. Just walk back and forth and you'll find it. Soon you'll come to the Winsor Trail...." He paused, allowing this possibility to sink in.

"And from there we could go to Spirit Lake and spend the night," I followed. I had backpacked to Sprit Lake once before via the Winsor Trail and knew how beautiful it was there. "Hmm...That sounds good."

To check what he had said, I raised the topographic map and looked for a trail on Penitente Peak. However, there was none shown. I looked back in his direction, but he was gone. He was gone as quickly and quietly as he first came upon us. "Where the heck did he come from?" I said to Janie, as much to myself. She didn't respond.

It came as a shock to me years later that apparently I was the only one to see him.—Janie had no recollection of the encounter. As surprising as it may be today, we never talked about it until decades later. There was nothing at the time that caused me to bring the subject up again. It was time to get to camp, and perhaps the fatigue of the day dampened our conversations. The only mention of this encounter occurred on the following day when I told a newspaper reporter that we had followed a primitive trail based on hearsay, meaning this encounter.

As a possible explanation as to why I saw him and apparently she did not, Janie suggested that maybe I'm more mystical than she is.

33

With little need for discussion, soon we were on top of Penitente Peak and saw a trail. Barely discernible at times, it was hardly wider than a boot and was otherwise unmarked. Though it wasn't clear what trail we were on, the path headed in the right direction so we followed.

As with summits on many of its neighbors, the top of Penitente Peak was above timberline. But walking on the top of Penitente was special. Its open mesa-like surface gave around-the-world views of the Rockies and the Rio Grande valley.

I felt liberated and slightly vulnerable at the same time. The emergence from the tree line brought expansive views that rewarded us. However, the landscape resembled Arctic Tundra with wind-swept dwarf evergreen bushes, and grasses sprinkled only with tiny forget-me-not flowers, and purple and yellow asters. The change in vegetation above timberline unmistakably stated that this was an environment with harsh winds and short growing seasons.

We hiked easily and followed the path. The gentle downhill slope and great views lightened our steps. As we approached the last of the open area, the trail took a swing to the east and entered the forest. The remaining late-afternoon sunlight, blocked by the dense tree canopy, failed to penetrate to the ground level and made the trail harder to track. The brilliant skies of the morning had now been subdued with clouds, typical of the daily weather cycle present in the high country. The dimness made it feel as if we were hiking by moonlight. There were places in the forest where the trail was completely covered with leaves and indistinguishable from the native soils.

We lost the trail, and then found it again, as predicted by the wool-clad man. The trail resumed its rough north-south direction and was aligned to intersect the east-west running Winsor Trail. I projected that we would come to the Winsor near a saddle in the notch between Penitente Peak and Santa Fe Baldy. We confidently hiked along.

Ten minutes later, the trail disappeared again—this time for good. Despite a zigzag search, we could not locate it. We were tired from the long climb and hike. Dark rainclouds had formed and it was starting to mist. Our energy levels had dropped and perhaps our judgment started to be clouded. This was our second time to lose the trail—for that matter, it appeared to have ended—and visibility was getting worse. It was approaching six p.m. and time was an issue because we still needed to reach Spirit Lake to camp. Maybe, we briefly discussed, we should just bushwhack and we'll soon intersect the Winsor Trail: Let's not waste anymore time looking for that silly excuse for a hiking trail. I had guessed that we only needed to drop maybe 300 feet to reach the saddle. So Janie pointed to the north and said with a smile and a determined voice, "Let's go that way!"

After we committed to enter the deep forest, the slope steepened dramatically. We immediately began to drop—three times faster than before and often at more than a 45 degree pitch. Sidestepping was often necessary because of the slope. We walked downhill expecting to encounter the Winsor Trail at any moment, yet we couldn't look far ahead because of the trees. Downward we went. Perhaps I had misinterpreted where we were on the map. We should have found the trail by now. Had we not been worn out, had the light been brighter, had it not been raining, we normally would have turned back at this point. With undeserved confidence or stubbornness, we continued to drop.

It finally crept into my head that maybe we were lost. Ordinarily, the thought of being lost gives me anxiety. This time, however, there was no such panic: We weren't really lost; we merely had lost the trail. After all, we had been warned by that gentleman that we might lose the trail. And this time I had a map and compass. I had hiked in the general area several times before and knew the key landmarks. We knew where we were heading, right?

This time I thought I had a mental picture of our location, and I was confident in our ability to get off the mountain and find the nearest hiking trail. Down we went as if we didn't know better, as if we were being pulled along.

34

The human eye is good at spotting movement. The eye can also be good at recognizing an object that is out of place, but it often takes longer for the senses to clarify. As we carefully made our way through the fallen trees, brush, and rock on the hillside, our gaze was focused on the ground immediately ahead. It was during one of the quick upward glances that an out-of-the-ordinary object was spotted; a piece of thin metal protruding from behind a group of trees. At first, it looked like an aluminum yard shed that had been crumbled by high winds. What was that doing here? A few more steps, however, and it soon became clear that this was an airplane!

Even after recognizing that we had stumbled across a plane crash site, it simply didn't register in my mind that it could have been a fairly recent crash. So I viewed the object as more of a curious thing and sat down to rest, while Janie went ahead to look at the plane. I awaited her report as to what was left of the wreckage. For a moment, she was out of sight as she circled around the line of trees to get a better view of the aircraft. In a matter of seconds, she backtracked and looked up to me. Then came her words that still give me goose bumps four decades later: "Bruce, there are bodies here."

I struggled to breathe. This was totally unexpected and unimaginable. This was something you don't prepare yourself for. I took a quick deep breath and hoped I could keep it together when I saw the bodies. It scared me to think I wouldn't do well. So probably out of self-preservation, I shifted my concern to how this will affect Janie. How will she take it?

The plane faced downhill. The tail of the fuselage hung in the evergreens about six feet off the ground. As I rounded the corner to view underneath the plane, I spotted skeletons. Skeletons! Thank goodness there were skeletons and not, well....bodies. This I can handle, I thought, as macabre as it was.

One skeleton lay immediately beneath the fuselage. It was intact from the skull through the pelvis. A second was less intact five feet away. Then were tens of large bone fragments and chips scattered around the wreckage in a 20 to 30 foot radius. It was obvious the animals had been active here. A pair of jeans lay on the ground. Along with the bone chips were personal effects: a binocular case, reading glasses, maps, and wallets. Bizarrely, a gasoline credit card stood upright on its edge, delicately held in position by short grasses.

To my untrained eye, at first the body of the plane looked to be in good shape. However, as we looked further it became clear that this was a violent collision. There were no seats in the aircraft. A piece of a seat belt on the ground appeared to have been ripped in half by the impact. A wing dangled 20 feet up in a tree. As we later scanned around the site, more evidence of extreme impact was seen. About 50 feet downhill from the nose was the engine, tightly wedged into a gap between two trees about 10 feet off the ground. Somehow the engine had become detached from the nose and flew like a bullet into the evergreens. The speculation by many was the ground was hit with such force that the nose of the plane rebounded off the underlying rock, and the engine detached as the plane flipped upward or backward. Eventually, the body of the plane settled, hidden beneath the tree canopy. I didn't see how anyone could live through the impact.

Janie and I were silent for several minutes as we looked around the site. Gradually, the initial shock of what we were seeing had passed. Janie turned to me. "What do you think?"

"Well, based on the skeletons, it looks like there were two people in the plane," I said. Janie nodded.

I continued. "And the crash wasn't too recent because the bones are clean. But my guess is the crash might have happened this year because of the way that credit card was still upright. With all the snow and wind this place gets, there is no way the card could stay like that for very long."

The light was getting dimmer, the rain was picking up, and we still had to get to camp. We needed to do something, but what?

I looked at Janie. "Why don't you go around and collect items that would help identify the people. I'll write down information about the plane. Then, I'll take some compass readings of where we are." It was a good sign when Janie sprang into action. I handed her a clear plastic bag and she began to collect wallets and the like.

I pulled out a pad of paper and a pencil from my pack and went to work looking for the plane's ID, not knowing a thing about aircraft IDs. I naturally started at the part of the plane which was closest to me—the tail. I glanced up and directly in front of me was a stainless steel placard, with all sorts of numbers, such as what appeared to be the plane's serial number. Bingo! This has got to be it! Immensely proud of myself for quickly finishing such a technical assignment, I returned to my pack and pulled out the compass. As I searched for a place to take a compass reading, I walked past the huge "N" numbers on the side of plane and completely ignored them.

Given the time of day and the age of the crash, we decided there was no point in trying to return home that night. Oh, and then there was the matter that we were still lost.

We hoisted our packs and continued downhill, carrying information that we were sure would change someone's life. In a half-hour, we bottomed out in a dry ravine, climbed up the side of the next hillside and finally found the Winsor Trail. With rain heavy, our ponchos came out of the packs and we dragged on to Spirit Lake.

Physically exhausted and mentally numb, I grabbed the tent and ground tarp. In no time the tent was standing, albeit a little droopy in the middle. I might have allowed more time to tighten up the middle section had it not been for the driving rain and fatigue. We had managed to survive a traumatic day. And, after a magnificent steak dinner, I thought, we could retire to a cozy structure and listen to the patter of rain.

Two solid hours of cold, torrential rain had soaked the landscape. Nonetheless, I was confident of getting a fire going because in the past I had attended my dad's fire building classes. One of those classes was held during the middle of a snow storm, when he challenged me to build a fire before he did. I quickly gathered the small kindling one would normally use in a fireplace. My dad instead slowly began to turn over downed timber and probe the wood with his hunting knife. Soon he dug out a piece of pitch, which provided the fuel to immediately start his fire. Mine, however, never reached passed the smoking phase. Just like this day. No matter what I used—pitch, fire starter, paper—I could not get a fire going in the driving rain. Eventually, I broke the bad news to Janie that we would not eat the steak tonight. She understood and we retreated to the tent to share crackers and peanut butter.

As weariness took over, we listened to the rain in the dark and allowed our bodies and minds to recover from the day. We dozed off in little time. Little did we know that as we slept the roof of the tent was slowly being deformed by the weight of accumulating water. A couple hours past midnight, with cold abruptness, the water found a weakness in the flaps and a roof-full of water rushed into the tent. Although our down sleeping bags stood in three inches of water, we huddled together in a high spot and waited for morning.

35

Sunday, August 5, 1973

I opened the front door of my parents' house around three p.m. and said hello to my dad in the entryway.

He asked, "How was the trip?"

"It was great." Knowing that my dad loved to identify plants and flowers in the wilds, I added, "The trail was covered by these tiny purple wildflowers."

"Some kind of aster, probably."

"Maybe," I offered. My dad turned and started to walk into the kitchen.

"Oh yeah. And we found an airplane crash."

My dad stopped, turned, and gave me a quizzical look, as if he hadn't heard correctly. My sister, who was sitting in the living room, jumped into the conversation with a loud "What?"

"And there were bodies there," I added, thinking that would clarify all. My dad looked flabbergasted, so I gave him a quick summary of what had happened.

"Well, you have to contact somebody," he urged.

"Who?"

"I don't know. Maybe the State Police."

Then, I remembered that one of my bosses flew airplanes and was involved with Civil Air Patrol searches. I reached him at home. Bob Abernathey listened to what we found and the line went silent.

"I think this is the Big One," he finally said.

"The Big One?"

"We looked all over the place for this and never found it. Mystery plane. It's been missing for about five years now."

"No way. I would have guessed one year," I said, thinking of the credit card still upright on its edge.

"Why don't you give me any information you have and I'll make a few calls," he offered.

I got out my sheet of paper and read the serial number and others to him.

"What about the "N" number?"

"The "N" number?"

"The big one on the side of the aircraft."

Oh, my. "I didn't write it down."

My boss called his CAP contacts. Unbeknownst to me, I had triggered an avalanche of activity. In a few hours, phone calls were placed to families across the country, and alerts were given to the State Police, the FAA, the military, and—oddly—to the news media.

Still not completely appreciating the magnitude of what we had found, I went upstairs to shower and got ready for dinner at Janie's house. The centerpiece of dinner was that massive steak. I was determined to eat what I had carried all over the countryside; the peril of eating warmed meat didn't register.

SUNDAY, AUGUST 5, 1973

Sara and the crew arrived back in Albuquerque at 4:30 in the afternoon following a family conquest of the southwest. In a day and a half they had visited El Paso, the Juarez tourist market, Carlsbad Caverns, and White Sands. After reaching Sara's house, they all needed some down time to swim and shower before their last stop.

To close out the road trip, the family piled into a few vehicles and headed to dinner and treats at Albuquerque's Old Town Plaza. Sara's mother stayed behind at the house to watch over the youngest child in the party, Andy, who desperately needed sleep.

The family sat around a large table at the Old Town restaurant and placed their orders for New Mexican food dishes. Then, they held hands, bowed their heads and prayed.

As the prayer ended, the phone rang fifteen miles away at Sara's house. Sara's mother answered. It was the FAA calling to say there was a strong likelihood the plane had been found. Mom was waiting in the driveway with the news when Sara and the others returned from dinner.

It was 1,768 days ago when her husband disappeared. Sara finally received word that the plane had been found. Word came as her immediate family gathered for the first time in five years.

ROCKLEDGE, FLORIDA, AUGUST 5, 1973

Don Arnett met Barbara Jones nearly four years ago. After the initial dating period, they rarely talked about Ron Jones or her first marriage. It had seemed that after Barbara had received what she called "the sign"—prior to her leaving Albuquerque—her life had moved forward. Don and Barbara had since married.

Don was talking with Barbara in the living room when the phone call came from Sara Rex. Sara had called her just two weeks before to say Hello, and to get another call so soon caused Barbara to think that something unusual had happened. As she heard that the plane had likely been found, Barbara had to steady herself and sat in the nearest chair.

She hung up the phone and began to weep uncontrollably. It was the first time that Don had seen her cry in those four years.

Although Barbara had mentally accepted Ron's death long ago, her reaction revealed the presence of much deeper emotions that perhaps she did not know existed. The news released hidden tension, anxiety, and love that had been suppressed for close to five years.

I had just started to enjoy the steak at Janie's house when my dad called.

"You had better get your tail up here. The phone is ringing off the hook. Police, newspapers—you name it.

"Okay. I'll be right there."

Twenty minutes later I walked in and was handed a long list of names to call *NOW*. I started with the New Mexico State Police. It was agreed that Janie and I would drive in the morning to the State Police Headquarters in Santa Fe and meet at nine o'clock with "a few people." Then maybe we would direct a search team to the wreckage.

I worked down the list. The FAA was contacted, followed by the

Albuquerque Journal, the *Albuquerque Tribune*, and others. I didn't recognize the last name on the list: Mr. Kenneth Rex. Although it was after 10 p.m., I called Mr. Rex.

"Mr. Rex. My name is Bruce Gallaher. I was told you wanted to talk with me."

"Bruce, thanks for calling. I represent the families of the men who might have been on the plane."

"I see." At his urging, I proceeded to tell him a condensed version of the discovery story, which had now been repeated many times.

"Well, it sure sounds like our plane. I just have one more question for you, to be certain: What was the "N" number on the side of the plane?"

I hesitated, and admitted that I did not know the "N" number. And then I blurted out, "But, I do have the serial number."

Silence. Then he continued with a very low and determined voice. "Bruce, this family has been through hell the last five years. For your sake, I hope this isn't a hoax."

"I swear to you, it isn't."

He seemed unconvinced. "Just last week, someone claimed that he had discovered a plane crash with four bodies in it—down south in the Sacramento Mountains. Well, they searched and he couldn't lead the police back to the plane." He closed the conversation with an ominous statement. "This better not be a copy cat of that. I'll be at the meeting in the morning and we'll see what you have. Good night."

HONOLULU, HAWAII, MONDAY, AUGUST 6, 1973

The phone rang about 2:00 in the morning in Honolulu. It was Beverly's mom on the line, and when her mom reported that they brought back Jon Dale's credit cards from the plane, she knew it wasn't a hoax. Beverly hung up the phone and didn't know whether to cry or collapse with relief. She had allowed herself to think about this moment many times before, but all that practice didn't seem to help. Nonetheless, her wish came true: Beverly would know soon what had happened to Jon Dale. She spent the rest of the night packing for the trip to Albuquerque while Douglas slept.

Word came for Beverly barely a day after she decided to give up the search and rely on fate.

MONDAY, AUGUST 6, 1973

My mom gently shook my shoulder at six in the morning. "Bruce, you had better get up and see something." My sleep was deep, yet short.

I dressed and wandered downstairs to the kitchen where dad and mom stood, coyly smiling.

"What is going on?"

My dad handed me the morning paper, the *Albuquerque Journal*. The Headline "Bodies Found Near Plane" knocked me over—huge letters of the kind normally reserved for assassinations or wars.

Albuquerque Journal, Monday morning, August 6, 1973

Bodies Found Near Plane
Cards Name Men Missing for 4 ½ Years
By SCOTT BEAVEN

Wreckage believed to be that of a plane carrying four Albuquerque men that has been missing for almost five years was discovered by a back-packer in the Pecos Wilderness Sunday.

Bruce Gallaher, 23, of Los Alamos, was hiking about one and one-half miles south of Santa Fe Baldy, a 12,622-foot mountain, when he took a wrong turn and discovered the wreckage believed to be the remains of a single-engine Piper Cherokee that left Albuquerque on Oct. 2, 1968 with a pilot and three men who were on an elk spotting trip in anticipation of the hunting season.

The four—John W. Fishel, 29, the pilot; John D. Horton, 27; Kenneth Brittain, 28, and Ronald P. Jones, 28—have been the subjects of extensive searches by government agents, private groups and even a clairvoyant physician.

Gallaher reported that two skeletons were near the plane. One was intact and the other was partially intact.

And he brought identification out of the wilderness with him bearing the name of John W. Fishel and other identification bearing the name J. D. Horton with the address ___, the family address at the time of the disappearance.

171

Gallaher said he did see any other skeletons and so, for the moment, the fate of Brittain and Jones is still a mystery.

State Police radio dispatcher Rudy Miller in Santa Fe said a ground party probably will be sent into the area today.

Civil Air Patrol wing commander Richard Damerow said that, if requested, he would send in a land team but that the order would have to come from Air Rescue in Kansas City.

Gallaher, an employee of Los Alamos Laboratories, told the Journal that he and a friend—Janie Hones, 23, also of Los Alamos—had gone into the wilderness for a weekend back-packing trip.

"We were on a primitive trail that is not on the topographical maps—we were going on hearsay, really, and we were trying to find a lake when all of a sudden the trail just petered out," Gallaher said.

"We kept walking and came up against what we thought was a hill but it turned out to the 3000-foot side of a mountain. We were zigzagging back and forth and halfway down the mountain we saw the fuselage of a plane. There was a tire in front and the propeller was there.

"There were no seats in the plane. One wing was hanging in a tree 40 feet west of the main body of the plane. We didn't see the other wing. The body itself was fairly intact with only slight creases on the sides. The major damage apparently was done to the wings and the front part of the plane.

"The nose was touching the ground and the tail was lodged in tree branches at a 45-degree angle to the nose. We walked up to the plane and beneath it was the almost perfectly preserved skeleton of a man."

Gallaher said, "Another skeleton was nearby. We saw some dungarees and looked through them and we found a wallet with a lot of identification for Horton. About 10 feet from that we found a shirt and another pair of pants and a shirt but there was no wallet. But 10 feet from that we found another wallet with some identification for John Fishel that stated he worked for Edgerton, Germeshausen and Grier Inc. of Albuquerque.

"We also found two watchbands near the first pair of pants and we found half of another watchband in front of the plane, indicating that there might have been more than two people.

"In the front of the plane, there were two pairs of shoes. As were leaving, we found of third pair of shoes—black safety boots.

Gallaher said he walked out of the wilderness Sunday afternoon and reported the findings to State Police about seven Sunday night. The plane was discovered about 6:15 Saturday evening, he said.

Gallaher wrote down all of the numbers he could find on the plane. Serial number 28-28-83, plane number MAA-Plate-317-671, Piper number PA-28-180.

Kenneth Rex, of Albuquerque, the husband of the former Mrs. Brittain said he did not know whether the numbers were those of the missing plane but he said he would check. He also said he would start calling the surviving relatives.

"This will be a relief to them if this is the real thing," he said. "It will be a relief to be about to put an end to all of the wondering and worrying."

He said Mrs. Horton lives in Nebraska and has not remarried. Mrs. Jones, who lives in Florida, has remarried. Mrs. Brittain married Rex and still lives in Albuquerque. Fishel was unmarried. The Fishel family, however, lives in Nebraska.

A $2000 reward has been offered for information leading to the discovery of the plane and Rex said the reward is still valid.

"Yes, sir, that reward is still for the person who finds that plane," Rex said.

When Gallaher was told about the reward, he was surprised and said, "Well, that is rather strange. Maybe it's a matter of fate or something."

The families of the four men have spent a great deal of money and time over the past few years searching for the plane.

In November 1968, the Air Force undertook a photo mission in northern New Mexico to try and locate the plane. Pilots—hired by the family and not—combed the area and, at one point, an Albuquerque physician believed to have powers of extra sensory perception tried to find the plane after he had been furnished with a wide range of pertinent data.

All efforts were fruitless and for nearly five years the families waited and wondered.

Precisely at seven a.m. the phone rang and my dad handed the receiver to me. It was the news director of 770-KOB AM radio in Albuquerque, a 50,000 watt giant of station that can be heard in 19 states.

"Joining us now live, is Bruce Gallaher."

Engine disassembled from fuselage and wedged between trees some 50 feet away.

Airplane propeller. Because the propeller was bent, inspectors concluded that the engine was operating before impact.

Airplane parts still in tree five years after crash.

One of the wives examining crash site.

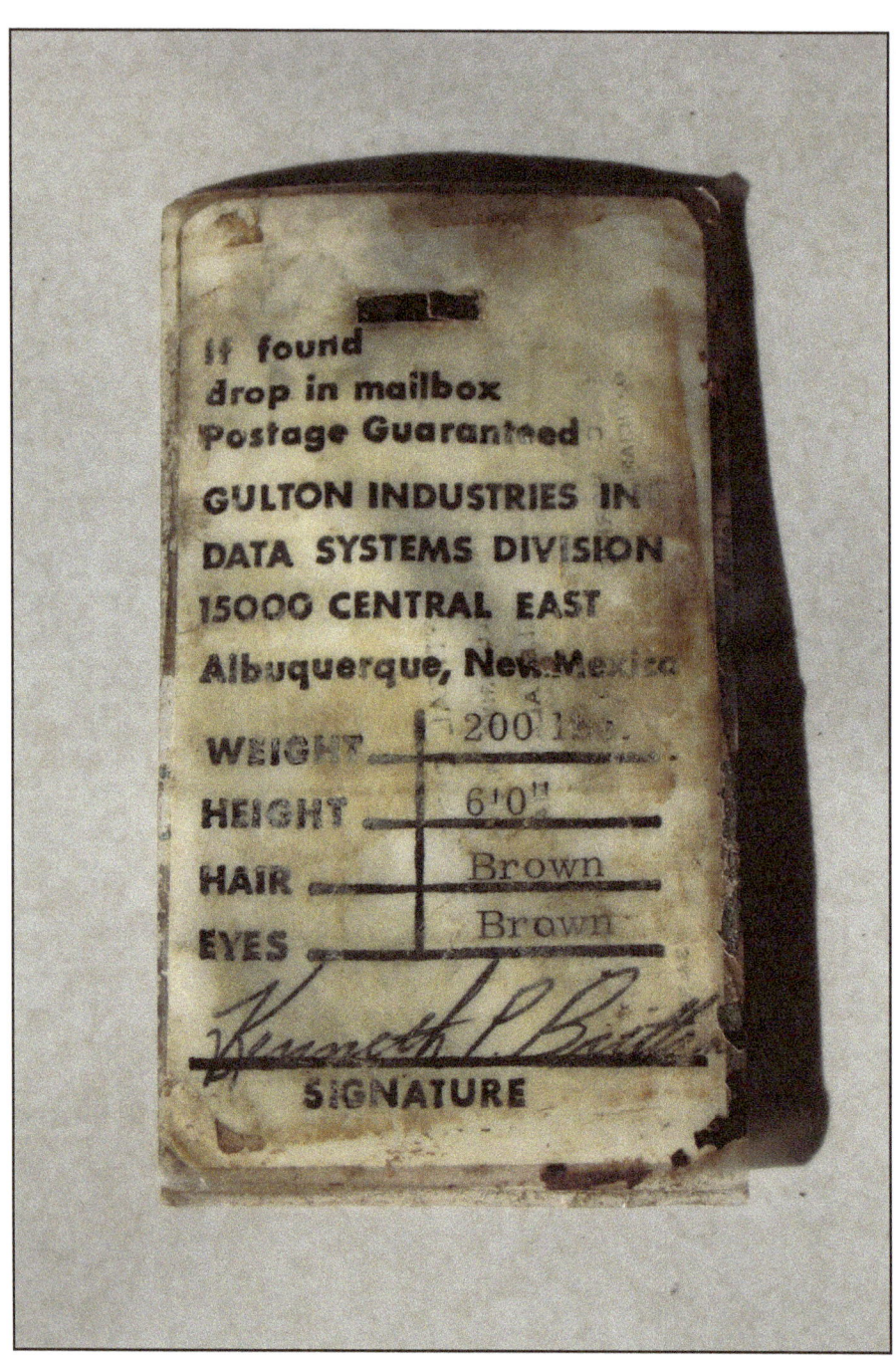

ID card of a passenger.

Personal effects of another passenger.

An FAA inspector with National Guard pilots in meadow near crash site.

36

At a few minutes before nine a.m., Janie and I walked into the headquarters of the New Mexico State Police, carrying the plastic bag full of wallets and other personal effects. A glistening new white building—the place spoke of authority. A receptionist was seated behind a long counter.

"Hi. We have an appointment with someone with the State Police. We found an airplane crash over the weekend and..."

"Oh, yes. They are waiting for you. It's the first door on the right down the hallway."

We opened the door expecting to see an officer or two and, of course, Mr. Kenneth Rex. Instead, we walked into a large conference room filled with about twenty news reporters. Along the back of the room hovered a ring of TV news cameras mounted on large tripods.

The Police Captain saw us and introduced himself. He smiled and asked us to come to the front of the room and sit at the empty table. After we settled, the officer turned and addressed the audience.

"Ladies and gentlemen. This is Bruce Gallaher and Janie Hones from Los Alamos. They have some information about the plane crash to share with you. Go ahead Bruce."

What? In a panic, I looked at Janie and she appeared to have a deer-in-headlight look. To address a room full of reporters. My hands began to shake, so I interlocked my fingers.

We were too petrified to contribute much. Besides, I figured that everybody in the room had already read the *Albuquerque Journal* article or heard the KOB radio interview. There was nothing more I could say, and I took the easy way out. In a very undignified move, I grabbed the plastic bag and poured its contents onto the table. I immediately regretted doing this, as I saw the names of the men on the plane inscribed on the personal effects. Looking over my shoulder when this happened was Kenneth "Jack" Rex, who slumped into a nearby chair.

The press conference wrapped up in about an hour and we were led into a back room. The large room was abuzz with activity, akin to what might take place before launching a military operation. People were on the telephone, others were gathering maps. A soldier dressed in green flight fatigues approached me and asked if I could provide directions to where the crash was located. After giving him general directions to Penitente Peak, I reached in my pocket and handed him the compass readings I had taken. The readings measured the bearings from the wreckage site to the summit of Santa Fe Baldy and two other prominent landmarks in the area. The soldier laid out the bearing lines on an aeronautical map and stared at the point where the lines converged—the location of the plane.

"Well, I'll be. I know exactly where this is," he said, while shaking his head in disbelief. The soldier went on to explain that he was a helicopter pilot and he would fly to the site if allowed. But first, he continued, they were waiting on permission from Washington, D.C. to land in a wilderness area. I knew that motorized vehicles were prohibited from the wilderness, except in life-threatening emergencies. What constitutes a "life or death" situation is subject to interpretation by the federal agencies. I wondered to myself, How in the world could they convince someone like the Secretary of Interior or the Forest Service chief that this was an emergency? The only official reference to this is in the final FAA report: "Special permission to land in the wilderness was obtained from the U. S. Regional Forester." The justification used remains unknown, but one hour later the pilot announced that they had a green light to fly and recover the bodies.

Janie and I were ready to go, when someone in the rescue team walked over to our table. He matter-of-factly explained that, because of weight restrictions, they could only take me on the helicopter. He then looked at Janie and asked her if she would be willing to hike back to the site, and lead some family members to the plane. There never was any discussion as to who was best qualified to be in the helicopter: it was decided ahead of time that, obviously, I would be that person. It was simply assumed that I was the reason the plane was found, and Janie must have played a secondary role, which could not have further from the truth.

Janie convinced her best friend, Sharon, to join her on the hike. They departed immediately for the Winsor Trail, accompanied by Jack Rex and members of Sara's extended family.

The National Guard helicopter was housed at the Santa Fe airport, only about a 15-minute drive from the State Police Headquarters. As I opened the door to my car, I was shocked to see a caravan of news media that had followed. I was becoming overloaded with all the attention. I identified myself to a clerk in the front, and he told me to wait in the hallway while they readied the helicopter.

"We are ready sir," the soldier said.

We walked down the hallway and out into a bright blast of sunlight. A hundred feet more and we arrived at the Army-green New Mexico National Guard Huey helicopter. The soldier led me to the left side of the aircraft. The other passengers already were in their seats. Along the opposite side of the helicopter were two members of the State Police, each dressed in shiny gray coveralls adorned with several patches, and matching ball caps. In the back, the FAA Inspector studied a notebook. The pilot and navigator gave a quick wave as I boarded.

I was asked to sit in the "hot seat" which faced out the large open side door. This was the seat with the best view and, accordingly, with added responsibility. The soldier then rigged me up with headphones and microphone for talking with the flight crew. Finally, the soldier took the adjacent seat, also facing the open door.

"Do you hear me okay, sir?"

"Yes," I answer.

Everyone was friendly, but there was no doubt that they were focused on the mission. There was little chatter, all business. They would get the Huey to the general vicinity and then it would be up to me. The main problem: Even with a helicopter, it would take a small miracle to spot it in the thick forest.

The entire state was watching this effort. Most importantly, though, I was mortified of letting down the families. I was deeply worried that I couldn't find it again, and I took a couple of deep breaths and just hoped. With little fanfare, the rotors started and soon we were airborne and headed East toward the mountains. The others on the Huey were seasoned fliers and they calmly relaxed in their seats. My senses, on the other hand, were exploding.

I had never been in a helicopter before and even the mundane looked different from the air. Although we were flying with the door open, the headphones largely dampened the sound. An ever-changing landscape slowly passed below, like a movie, as we gained altitude and soon entered Wilderness. As I looked out the door, my anxiety increased. Nothing looked familiar to me. Before I recognized any landmarks to help orient myself, the headphones came alive.

"We are approaching the area, Sir", the pilot said.

"Okay", as I struggled to figure out where we are. A half minute passed.

"Just let us know when you think we are close, Sir."

"Okay." Silence. I didn't look at the others in the helicopter, but I sensed their stares.

"I hate to say this, but nothing looks familiar," I said. "Can we fly over the meadow at Puerto Nambe, so I can get my bearings?"

Finally, as we hovered at Puerto Nambe—the saddle at the notch between Penitente Peak and Santa Fe Baldy—the landscape looked familiar!

"Now, let's double back," I told the pilot.

The Huey banked and we returned to where the pilot first alerted me. We flew slowly at maybe 150 feet above the ground, tracing the steep slope of the mountain. My eyes scanned the few openings between the trees. The forest mostly resembled a continuous dark carpet from the air. Visibility was reduced because we were partially blinded by the mid-morning sun that had recently risen over the peak, creating a patchwork of brightness in the few open areas and deep darkness in most of the forest.

Suddenly, I felt confident. It just felt right. Although I had seen nothing of the plane, I pushed the button to talk into the headset.

"If my memory serves me, the plane should be about right there," as I pointed toward the hillside.

Less than a second later, an intense flash of light reflected from the ground.

"I see it! I see it!" someone on the helicopter yelled. "I can lead us there."

I knew that the plane's fuselage and engine were hidden beneath a nearly complete tree canopy. At the very instant I pointed, against all odds, we happened to be flying at the precise altitude and time of day to catch a reflection off a random piece of metal debris scattered to a small opening in the tree canopy on a steep north-facing slope.

The helicopter flew back to the large meadow and began to descend. Directly below us, two young chubby backpackers, possibly about 16 years in age, looked up and saw a U.S. Army helicopter coming toward them in the Wilderness. Their eyes grew large and, without discussing their options, took off running with loaded packs bouncing hugely up and down. We never saw them again, making one wonder what they had in their packs.

What a moment! The intense pressure lifted from me. I had done my job. While we awaited the flight crew and the specialists to get organized for the hike to the plane, the pilot pulled me aside.

"That is amazing," the pilot said. "You know, years ago I was involved with the search for that plane. I had flown that identical route at least three other times. We saw nothing."

After a few seconds of thought, he turned back in my direction and continued. "You are either one lucky guy, or there was a miracle here."

His words caused me think of the mountain man on the trail, along with the flash of light. I wondered.

The soldier was true to his word. He led the team right back to where the brilliant flash of light was seen.

The 'recovery operation' was surreal. There was little talking between the team members, who efficiently and professionally went about their jobs. The FAA inspector headed for the propeller to see if it was rotating on impact. A person scoured the ground and bushes looking for hidden clues or parts. Another group readied body bags. Time seemed to slow down, as though each person was caught in their own bubble of thought.

Among the items recovered was a book "A Hiker's Guide to Pecos Wilderness." that belonged to Jon Dale Horton. The book had a place marker in the page for Penitente Peak.

I was the only one without a specific job to do. Eventually, I walked over to the FAA inspector who was finishing up his notes. I asked him what he had found. He prefaced his answer with a "This is preliminary" statement, and went on to visualize the crash for me. He pointed to the top of a line of tall evergreens to the west of the fuselage; their tops had been snapped off and a wing flap was hanging in a tree approximately 20 feet above the ground. Clearly, the plane had hit the trees while traveling in an easterly direction and then impacted the ground. An intact propeller for the aircraft was found. One blade of the propeller was buried in the ground. The aircraft fuselage was approximately 50 feet from the propeller. The fuselage, lying on its left side, was demolished, in stark contrast to what I had initially told the press based on my untrained eye.

Then he summed things up by speculating that the plane probably was flying up canyon, to the west and towards the saddle at Puerto Nambe, between Penitente Peak and Santa Fe Baldy. The plane likely couldn't overcome the downdrafts encountered on the leeward side of the saddle. Losing altitude quickly, the pilot probably banked left to reverse course but couldn't quite clear the trees. The official FAA report gave further detail:

"....The front of the cabin section—forward of the instrument panel—had broken away from the fuselage. The front seats were missing. It appeared that the aircraft damage was the result of the trees and ground impact while at a high rate of speed. There was no evidence of fire.

"The engine was approximately 150 ft. down slope from the fuselage. Attached to the engine was the instrument panel. No readings with the exception of the tachometer were obtainable. The front seats were found approximately 75 ft. down slope from the engine. All structural components for the aircraft were identified from pieces or parts of the wreckage. A check of the aircraft systems could not be made due to the extent of the damage.

"In summary: It appeared, from the damage to the tree tops that may have been the result of the aircraft striking them, and the wreckage distribution, that the pilot was in a turn when a wing, or part thereof, struck one or more trees."

In the official FAA report, the cause of the crash was the encounter with the trees; few of the other aspects of the scenario discussed at the site appeared in the report—for example, downdrafts near the saddle—because they were speculative and couldn't be verified (this plane was not equipped with a 'black box' which could be used for accident reconstruction).

I don't know how long we were at the wreckage for the recovery of bodies; probably no more than two hours. Just when we were ready to return to the helicopter, Janie, Jack Rex and the ground party arrived at the site, much to my admiration. Years afterward, the ground hikers spoke in awe of Janie's strength and endurance.

Back in Albuquerque, family members anxiously awaited word of the rescue operation that had taken place in the Pecos Wilderness. Indeed, *their* plane was found and soon the nightmare would be over. First, however, the bodies of the men in the plane must be identified. Sara got dental X-rays of Ken Brittain from his doctor in anticipation of this step.

Late that afternoon, Ken Brittain's parents arrived from Phoenix. And then Beverly Horton flew in from Hawaii. The next day, Jon Dale Horton's parents arrived from Dallas, as did John Fishel's parents from Nebraska. Each family was asked by the Office of the Medical Investigator to identify personal belongings found at the crash site. Additional bone fragments found at the site by Jack Rex were submitted to the OMI.

Thursday August 8, 1973

Sara was joined by eleven members of her and Ken's families for the hike to site to view the wreckage. They were accompanied by Beverly and the Horton parents. Unfortunately, the Horton parents were unable to complete the hike because of the initial steepness of the trail. For the wives, the crash didn't seem to register at first. They struggled to attach themselves to the twisted parts of the aircraft. In the two hours they spent at the site, their emotions fluctuated from relief, exhaustion, to anger—or maybe it was disappointment.

The wives were disappointed that others had gone through the wreckage before they did. They felt that the authorities could have waited. What was the rush to remove the remains after all this time? In a sense, they believed this place should be treated as though it was a sacred site until the families had a chance to re-connect with their loved ones. As an indication of the importance of the site, the families asked me not to disclose the specific location of the wreckage.

That evening, Janie and I drove to Sara's house to receive the $2000 reward. It would be our first time to meet Sara or Beverly and we were nervous. I did not know how I should act. Should I be reserved in honor of the families? Or, happy because their agony is over?

Although it had been a trying day for the wives, they greeted us at the door with smiles and hugs. We immediately felt at home, as if we were family. A newspaper photographer was there to document the event. It was the last time in 35 years I was to see the wives.

37

The discovery of the missing plane changed many lives. Most importantly, for the families and friends of the men in the plane it brought closure.

As I came to learn over the years, this was a story of unyielding determination and love. This was a story of families that pushed themselves, government authorities, and even psychics to find the answer. This was a story of surprising corporate humanity and tenderness.

The effort ranks among the largest searches in the state's history. Key elements included the following:

> The 422 CAP personnel, 211 aircraft, and 538 flight hours in October, 1968;
> the F-4 photo reconnaissance;
> Doug Rhinehart and his stunt plane;
> the U-2 flights and photo interpretation;
> the numerous ground searches, involving more than 500 volunteers;
> the reward funds;
> the extensive media coverage;
> the postings and air drops of flyers;
> the 1100 flight hours by volunteer private pilots;
> the psychic readings of Jeane Dixon, Dr. Gilbert Holloway, and Peter Hurkos;
> the hired helicopter flights.

So why was so much effort given to help find four men—four seemingly regular guys? Why did these four men, as decent as they were, inspire this level of attention?

Possible explanations range from the exotic to the simple. On the exotic side, one explanation that has been offered is that two of the men

(Ron Jones and Jon Dale Horton) had worked at one of the nation's most secret facilities in Nevada and on development of the SR-71 spy plane. Could it be that such an effort was directed by the Government because these men held deeply-guarded secrets? This seems highly unlikely. There is no evidence from interviews or records to support this theory.

As best as the author can tell, the search grew to this magnitude simply because of passion. Numerous groups and individuals voluntarily were involved in one way or the other. Each selflessly contributed energy and passion to the search as they could. However, it appears it was the sweet synergy of these separate efforts that yielded these results.

Consider the Civil Air Patrol. There was no shortage of accidents for them to be involved with in 1968. However, in nearly every case, missing planes were located within one or two days. This mission proved to be far more difficult because it involved high-altitude wilderness terrain with thick forest. The CAP worked diligently day by day, rotating volunteers and aircraft. In the end, they searched for 11 days, and with no new leads and fatigue of searchers, they called off the effort. More than half of their mission flying hours for the entire year were used on this single search. Without doubt the CAP search would have continued if new solid leads had developed. All of the CAP personnel loved flying, but it was their passion for helping their neighbors that caused them to not easily give up on the mission.

It would have been easy for Ed Whaley and the managers of Gulton Industries to stand aside and assume that the authorities would take care of the search. But, this involved their people. The company was small enough that it felt as if some of their own family members were missing. Initially, the focus of Whaley and others was on helping the CAP to quickly locate the missing men. After it appeared unlikely the men would be found alive, however, they didn't slow down their search. The company's humanitarian side came to light when they shifted their concern to those of the wives. Whaley worried how the wives would make it until survivor benefits were paid by the insurance companies. As a start, the company continued to pay the men's regular salaries for three months until some insurance monies came in. By allowing staff to participate in searches during work hours, Whaley and the company made a decision that humanity was more important at the moment than corporate business. Many of the volunteer flying hours were performed by Gulton staff, and the company was instrumental in sustaining a coordinated search effort.

For members of the Eastern Hills Baptist Church, their love and respect for Sara and Ken Brittain drove them to be a major force in the search. They comprised a majority of the ground searchers, and led the way in putting together the initial reward fund. Some members of the church took to writing letters to agency officials and the congressional delegation, maximizing the government's interest in the case. The church's involvement came from within—separate from CAP or Gulton activities.

The media indirectly played a large role in sustaining the public's interest in the search. Because all four men were from Albuquerque, the coverage was probably more extensive than if they were from a more rural location in the state. Every news story brought more leads. It was an article in the *Albuquerque Journal* that motivated Doug Rhinehart to offer his time and stunt plane for the search.

Keeping the search alive for four years were the wives and parents. Although few in numbers, their influence was everywhere. It likely was either a "long shot" letter or a telegram from Jon Dale Horton's parents or a chance political connection they had that convinced President Johnson to approve the U-2 flights. Donations given to Mr. Horton by his co-workers helped pay for hired helicopter flights. The wives pursued every possibility to affect the search. While the plane's disappearance took innocence from them and disrupted their lives, the wives were driven to find answers and kept going with help from a hidden reservoir of determination.

38

For Janie and me, our 15 minutes of fame had stretched into at least one week. With stories appearing daily in the newspapers and an occasional mention on television, we became fixtures in many homes. As the attention faded with the story, I welcomed the pause and returned to work. By no means was the attention over, however.

I returned home that evening and received a phone call from a stranger. The conversation haunted me:

"Mr. Gallaher, my name is ___ and I live in Colorado. I read about the plane crash you found and wanted to talk to you."

"What can I do for you?"

"I was hoping you'd be willing to help me find my husband," she spoke softly. "You see, his plane has been missing for seven years now and it seems as if you have a gift for finding missing planes. I really need help."

As delicately as I could think of to respond, I told her that I didn't think I had any special gift or powers. I wouldn't be of any help to her. She listened quietly and then began to cry.

"Thanks anyhow," she finally said. "I was just hoping. I never give up, and I latch on to anything new that might find an answer."

Epilogue

Sara Rex (Brittain)

Sara and her husband Jack moved to central Texas to be closer to his mother. Sara continues to play the piano and organ for church. She believes the experience of the plane crash brought a positive lasting change to her life. When her friends are distressed, she uses the experience to teach them that good things can come from bad. Jack still roots for the Dallas Cowboys.

Sara's sons think of Jack as their Dad. As adults, they developed an increasing interest in learning more about Ken Brittain, their biological father. They came to visit in 2008 and learn about my research. In honor of Ken, the boys hiked into the Pecos Wilderness and visited the wreckage site on the 40th anniversary of the plane crash.

Beverly Horton

Beverly taught art in Nebraska for nine years. The long winters and her mother's illness brought her back to Texas in 1980, where she taught at Cooper High School in Abilene for 20 years. Through a parents without partners group, she met and developed a friendship with Jay Norman, and discovered they shared a love of dancing and of boating. Beverly and Jay married in 1983, fourteen years after Jon Dale's disappearance. Douglas was 13 and got an automatic little brother of 12, Jay Jr., plus a life-long friendship.

After retiring from teaching, Beverly and Jay relocated in 2000 to the Albuquerque area. Her house is colorful and fun, as is her personality. Her passions are swimming and displaying her '66 Landau Ford T-bird at classic automobile shows; Jay is rebuilding a '55 F100 Ford pickup. Even 40 years after the loss of Jon Dale, she remains quite emotional when she is asked to look back in time. She rebounds quickly, however, and soon laughs loudly with self-deprecating humor.

Douglas is married and works for the US Civil Service. He currently lives with his family near Tokyo, Japan. He has three children.

Barbara Jones

Barbara lived on the eastern coast of Florida with her second husband, Don Arnett, until her death in 2006. Her children, her church and Don remained the center of her life for decades. She did missionary work with children for over 20 years in Haiti and in the Bahamas. Barbara and Don's children are scattered around the Midwest and Southwest. They visit Don regularly and often see Barbara's parents and in-laws. Don guiltily admits that the plane crash was the best thing ever to happen to him personally. Without the crash, Barbara wouldn't have returned to Florida and they wouldn't have met.

The Fishel Family

Mr. and Mrs. Fishel remained in North Bend, Nebraska until their passing. According to John's sister, Betty, Mrs. Fishel was never the same after the plane crash, and her health declined. Mr. Fishel dedicated much of his spare time tinkering and maintaining John's last car, the Pontiac GTO with a special paint job. Betty Aldrich was the last surviving member of the family at the time of this writing; her sister died in 2007. Since the plane crash, her life was a mix of triumph and tragedy, yet she emphasized the positives. For most of her adult life, she was a long-time owner of a ballroom dance studio. Her husband died at age 55 in 1985 and her fiancé died in 1988 at age 52. "So then I gave up! But I have a son & daughter...both married with families, so I have much to live for," she wrote.

Janie Hones

Janie used her share of the reward funds to travel to Sao Paolo, Brazil and sing opera. She returned to the Los Alamos area within a year, and met her future husband. She remains an avid hiker and was a major driving force in establishing an elaborate network of hiking trails in the area.

Doug Rhinehart

Doug Rhinehart continued to manufacture, restore and fly Rose Parrakeets until his death in 1978. He died several days after he crashed after takeoff from Aztec, New Mexico. Beverly Horton was notified by friends that he was seriously injured, and was told Rhinehart had caught his landing gear in some electrical lines. The last of several air shows and fly-ins at Alameda Airport was the Walt Hall/ Doug Rhinehart Memorial Fly-in, held in 1979 by the New Mexico Antique Airplane Association. The NM AAA was formed in 1968 by Doug Rhinehart.

His youngest son, Dan Rhinehart, completed a full restoration of his father's Rose Parrakeet acrobatic show plane in 2005, and flew it for the first time in 20 years. He and other family members are working to restore other Parrakeets.

Ed Whaley, Gulton Industries

After 37 years with the company, Ed retired in 1996 and currently lives in Placitas, NM. Looking back, he is especially proud of the company's contribution to the Apollo Space Program and Gulton's role in developing the command and data handling subsystems for the global positioning satellites. Their first trip after retirement was a safari; his wife of 47 years passed away in 2005. Ed became a great supporter of this book and encouraged me to write. He still doesn't believe psychics should get paid unless they are successful.

Bill Carpenter

Bill eventually opened his own insurance company in the Albuquerque area. After 20 some years in the business, he has never had to deal with another missing person case.

George Friberg

George worked for Gulton Industries Inc. with Ed Whaley in Pennsylvania and New Mexico. George left Gulton Industries in 1986, and he currently works for Technology Venture Corporation (TVC). TVC is one of the nation's leading organizations in identifying breakthrough technologies in the government laboratories and transferring that technology to the private sector of the economy. George and his wife, Mary, reside in Albuquerque, New Mexico.

Jeane Dixon, Psychic

Ms. Dixon regularly provided psychic "intelligence" to President Nixon, who called her "the soothsayer". In one of the most bizarre situations, Nixon placed the country on alert after Dixon warned his secretary that a terrorist attack on America was inevitable. Dixon was also one of several astrologers who gave advice to Nancy Reagan during the presidency of Ronald Reagan.
Dixon died in 1997, aged ninety-three.

Peter Hurkos

Hurkos remained one of the country's highest profile psychics until his death in 1988. Although he was featured regularly on special television programs, his work was controversial to the end.

Dr. Gilbert Holloway

A student of Dr. Holloway gave him 80 acres of land in Deming, New Mexico. He and his wife moved there from Miami, Florida and developed the *Christ Light Community Church* into an "international center of spiritual healing and metaphysical Truth teaching."
Holloway passed away in the late 1990s.

Suggested Reading

This book is largely based on a collection of detailed notes and personal memories. To include all the references would fill many pages. The following list of books and magazine articles is useful for readers interested in learning more about key topics mentioned in the text.

Civil Air Patrol. *CAP Mission Aircrew Observer Course, Observer Training Slides 8842C7E503209*, Civil Air Patrol, 2008.

Desmarais, John William. *Search Planning Guidance For Use in General Aviation Missing Aircraft Searches in the Continental United States*, National Headquarters: Civil Air Patrol, 2000.

Elman, Steve and Alan Tolz. *Burning Up The Air: Jerry Williams, Talk Radio, and the Life In Between*, Beverly, Massachusetts: Commonwealth Editions, 2008.

Holloway, Gilbert. *E.S.P. and Your Super-Conscious*, Louisville, Kentucky: Best Books Inc. 1966.

1960-1970 National Transportation Safety Board. *Annual Review of Aircraft Accident Data: U.S. General Aviation, Calendar Year 1970*, NTSB/ARG-74-1, Washington,DC, Table 171.

Hoversten, Paul. "Diary of a Spy", *Air and Space Smithsonian*, Vol. 27, No. 1, (2012), 30-35.

Julyan, Robert. *The Mountains of New Mexico*, Albuquerque: The University of New Mexico Press, 2006.

Keen, Capt. Rangi. "Civil Air Patrol: The Official USAF Auxiliary," http://www.lebanoncap.org/resources/presentations/search-and-rescue-ppt.

Lerner, Preston. "Wingman in a Pontiac", *Air and Space Smithsonian*, Vol. 27, No. 1, (2012), 36-37.

Kelley, Vincent C. *Albuquerque Its Mountains, Valley, and Volcanoes, Scenic Trips to the Geologic Past No. 9*, Socorro: New Mexico Bureau of Mines and Mineral Resources, 1974.

Pearce, T.M., ed., *New Mexico Place Names*, The University of New Mexico Press, 1965.

Randi, James. *Flim Flam!*, Buffalo, New York: Prometheus Books, 1986.

Southwest Natural and Cultural Heritage Association. *Pecos Wilderness Santa Fe and Carson National Forests*, 1991.

Vittitoe, Charles N. "Did High Altitude EMP Cause the Streetlight Incident?" Albuquerque: Sandia National Laboratories Report System Design and Assessment Notes, Note 31, 1989.

Acknowledgements

This book would not have been possible had Sara and Jack Rex not kept search records for over 40 years. Also, it would not have been possible if Ed Whaley not kept such detailed notes, as he was accustomed to doing in his engineering practice. Lastly, without Sara's blessing for this project, I likely would have just walked away from it. She graciously agreed to make available critical entries in her personal diary that vividly described her ups and downs during the first months of searching. Sara took it on herself to connect me with the other wives.

I simply cannot thank enough Sara Rex, Beverly Norman, Don Arnett, and Betty Aldrich for bravely sharing their memories. It had been such a long time since they thought about this period, and the recollections were sometimes painful. Others who were critical in developing this story include Tony Arnold, Bill Carpenter, George Friberg, Joe Hoffman, Janie Hones, Dick Reiff, Jack Rex and Ed Whaley. James Norbit and Bob Abernathey kindly provided me with a primer on Civil Air Patrol training and search procedures. Felicia Lujan was helpful in gathering information at the New Mexico State Records Center and Archives. Thanks go to Fred Gallaher, Beverly Norman, Barb and Steve Pratt, Sara Rex, and Ed Whaley who volunteered their time to read early drafts and offered extremely helpful corrections and suggestions on ways the manuscript could be improved. My wife Maggi gave me unconditional support for this project and was my biggest fan while I charged ahead and chased spirits of the past. Finally, my belief in the good of humanity was bolstered by the many heroic acts by employees of Gulton Industries and EG&G, the volunteers at the CAP, and the devoted friends of Sara and Ken Brittain at the Eastern Hills Baptist Church.

www.ingramcontent.com/pod-product-compliance
Lightning Source LLC
Chambersburg PA
CBHW051126160426
43195CB00014B/2361